**Computer Vision for Structural Dynamics
and Health Monitoring**

Wiley-ASME Press Series

Computer Vision for Structural Dynamics and
Health Monitoring
Dongming Feng, Maria Q Feng

Theory of Solid-Propellant Nonsteady
Combustion
Vasily B. Novozhilov, Boris V. Novozhilov

Introduction to Plastics Engineering
Vijay K. Stokes

Fundamentals of Heat Engines: Reciprocating
and Gas Turbine Internal Combustion Engines
Jamil Ghojel

Offshore Compliant Platforms: Analysis, Design,
and Experimental Studies
Srinivasan Chandrasekaran, R. Nagavinothini

Computer Aided Design and Manufacturing
Zhuming Bi, Xiaoqin Wang

Pumps and Compressors
Marc Borremans

Corrosion and Materials in Hydrocarbon
Production: A Compendium of Operational and
Engineering Aspects
Bijan Kermani and Don Harrop

Design and Analysis of Centrifugal Compressors
Rene Van den Braembussche

Case Studies in Fluid Mechanics with
Sensitivities to Governing Variables
M. Kemal Atesmen

The Monte Carlo Ray-Trace Method in Radiation
Heat Transfer and Applied Optics
J. Robert Mahan

Dynamics of Particles and Rigid Bodies: A
Self-Learning Approach
Mohammed F. Daqaq

Primer on Engineering Standards, Expanded
Textbook Edition
Maan H. Jawad and Owen R. Greulich

Engineering Optimization: Applications,
Methods and Analysis
R. Russell Rhinehart

Compact Heat Exchangers: Analysis, Design and
Optimization using FEM and CFD Approach
C. Ranganayakulu and Kankanhalli N.
Seetharamu

Robust Adaptive Control for Fractional-Order
Systems with Disturbance and Saturation
Mou Chen, Shuyi Shao, and Peng Shi

Robot Manipulator Redundancy Resolution
Yunong Zhang and Long Jin

Stress in ASME Pressure Vessels, Boilers, and
Nuclear Components
Maan H. Jawad

Combined Cooling, Heating, and Power Systems:
Modeling, Optimization, and Operation
Yang Shi, Mingxi Liu, and Fang Fang

Applications of Mathematical Heat Transfer and
Fluid Flow Models in Engineering and
Medicine
Abram S. Dorfman

Bioprocessing Piping and Equipment Design: A
Companion Guide for the ASME BPE Standard
William M. (Bill) Huitt

Nonlinear Regression Modeling for Engineering
Applications: Modeling, Model Validation,
and Enabling Design of Experiments
R. Russell Rhinehart

Geothermal Heat Pump and Heat Engine
Systems: Theory and Practice
Andrew D. Chiasson

Fundamentals of Mechanical Vibrations
Liang-Wu Cai

Introduction to Dynamics and Control in
Mechanical Engineering Systems
Cho W.S. To

Computer Vision for Structural Dynamics and Health Monitoring

Dongming Feng
Senior Engineer, Thornton Tomasetti
NY, USA

Maria Q. Feng
Renwick Professor, Columbia University
NY, USA

This Work is a co-publication between John Wiley & Sons Ltd and ASME Press

Registered Offices
John Wiley & Sons, Inc., 111 River Street, Hoboken, NJ 07030, USA
John Wiley & Sons Ltd, The Atrium, Southern Gate, Chichester, West Sussex, PO19 8SQ, UK

Editorial Office
The Atrium, Southern Gate, Chichester, West Sussex, PO19 8SQ, UK

For details of our global editorial offices, customer services, and more information about Wiley products visit us at www.wiley.com.

Wiley also publishes its books in a variety of electronic formats and by print-on-demand. Some content that appears in standard print versions of this book may not be available in other formats.

Library of Congress Cataloging-in-Publication Data

Names: Feng, Dongming, 1985– author. | Feng, Maria, 1961– author.
Title: Computer vision for structural dynamics and health monitoring /
   Dongming Feng, Ph.D, Professor, Southeast University, Maria
   Feng, Ph.D, Renwick Professor, Columbia University.
Description: First edition. | Hoboken, NJ : John Wiley & Sons, Inc., [2021]
   | Series: Wiley-ASME press series | Includes bibliographical references and index.
Identifiers: LCCN 2020017110 (print) | LCCN 2020017111 (ebook) | ISBN
   9781119566588 (cloth) | ISBN 9781119566564 (adobe pdf) | ISBN
   9781119566571 (epub)
Subjects: LCSH: Structural dynamics–Data processing. | Structural health
   monitoring–Data processing. | Computer vision–Industrial applications.
Classification: LCC TA654 .F46 2020 (print) | LCC TA654 (ebook) | DDC
   624.1/710285637–dc23
LC record available at https://lccn.loc.gov/2020017110
LC ebook record available at https://lccn.loc.gov/2020017111

Cover Design: Wiley
Cover Images: Golden Gate Bridge, San Francisco
© ventdusud /Shutterstock, eye with visual effects
© SFIO CRACHO/Shutterstock

Set in 9.5/12.5pt STIXTwoText by SPi Global, Pondicherry, India

Printed and bound by CPI Group (UK) Ltd, Croydon, CR0 4YY

10 9 8 7 6 5 4 3 2 1

Contents

List of Figures *ix*
List of Tables *xv*
Series Preface *xvii*
Preface *xix*
About the Companion Website *xxi*

1   Introduction *1*
1.1   Structural Health Monitoring: A Quick Review *1*
1.2   Computer Vision Sensors for Structural Health Monitoring *3*
1.3   Organization of the Book *7*

2   Development of a Computer Vision Sensor for Structural Displacement Measurement *11*
2.1   Vision Sensor System Hardware *11*
2.2   Vision Sensor System Software: Template-Matching Techniques *15*
2.2.1   Area-Based Template Matching *16*
2.2.2   Feature-Based Template Matching *20*
2.3   Coordinate Conversion and Scaling Factors *22*
2.3.1   Camera Calibration Method *23*
2.3.2   Practical Calibration Method *25*
2.4   Representative Template Matching Algorithms *28*
2.4.1   Intensity-Based UCC Technique *28*
2.4.2   Gradient-Based Robust OCM Technique *33*
2.4.3   Vision Sensor Software Package and Operation *39*
2.5   Summary *40*

3 **Performance Evaluation Through Laboratory and Field Tests** *43*

3.1 Seismic Shaking Table Test *43*

3.2 Shaking Table Test of Frame Structure 1 *46*

3.2.1 Test Description *46*

3.2.2 Subpixel Resolution *47*

3.2.3 Performance When Tracking Artificial Targets *48*

3.2.4 Performance When Tracking Natural Targets *49*

3.2.5 Error Quantification *51*

3.2.6 Evaluation of OCM and UCC Robustness *51*

3.3 Seismic Shaking Table Test of Frame Structure 2 *56*

3.4 Free Vibration Test of a Beam Structure *59*

3.4.1 Test Description *59*

3.4.2 Evaluation of the Practical Calibration Method *60*

3.5 Field Test of a Pedestrian Bridge *63*

3.6 Field Test of a Highway Bridge *66*

3.7 Field Test of Two Railway Bridges *67*

3.7.1 Test Description *69*

3.7.2 Daytime Measurements *72*

3.7.3 Nighttime Measurements *72*

3.7.4 Field Performance Evaluation *75*

3.8 Remote Measurement of the Vincent Thomas Bridge *81*

3.9 Remote Measurement of the Manhattan Bridge *82*

3.10 Summary *87*

4 **Application in Modal Analysis, Model Updating, and Damage Detection** *89*

4.1 Experimental Modal Analysis *91*

4.1.1 Modal Analysis of a Frame *91*

4.1.2 Modal Analysis of a Beam *97*

4.2 Model Updating as a Frequency-Domain Optimization Problem *101*

4.3 Damage Detection *108*

4.3.1 Mode Shape Curvature-Based Damage Index *108*

4.3.2 Test Description *109*

4.3.3 Damage Detection Results *110*

4.4 Summary *112*

5 **Application in Model Updating of Railway Bridges under Trainloads** *115*

5.1 Field Measurement of Bridge Displacement under Trainloads *116*

5.2 Formulation of the Finite Element Model *118*

5.2.1 Modeling the Train-Track-Bridge Interaction *118*

5.2.2   Finite Element Model of the Railway Bridge   *120*
5.3   Sensitivity Analysis and Finite Element Model Updating   *121*
5.3.1   Model Updating as a Time-Domain Optimization Problem   *122*
5.3.2   Sensitivity Analysis of Displacement and Acceleration Responses   *123*
5.3.3   Finite Element Model Updating   *127*
5.4   Dynamic Characteristics of Short-Span Bridges under Trainloads   *130*
5.5   Summary   *136*

6   Application in Simultaneously Identifying Structural Parameters and Excitation Forces   *139*
6.1   Simultaneous Identification Using Vision-Based Displacement Measurements   *140*
6.1.1   Structural Parameter Identification as a Time-Domain Optimization Problem   *141*
6.1.2   Force Identification Based on Structural Displacement Measurements   *142*
6.1.3   Simultaneous Identification Procedure   *144*
6.2   Numerical Example   *146*
6.2.1   Robustness to Noise and Number of Sensors   *147*
6.2.2   Robustness to Initial Stiffness Values   *150*
6.2.3   Robustness to Damping Ratio Values   *150*
6.3   Experimental Validation   *154*
6.3.1   Test Description   *154*
6.3.2   Identification Results   *155*
6.4   Summary   *157*

7   Application in Estimating Cable Force   *171*
7.1   Vision Sensor for Estimating Cable Force   *172*
7.1.1   Vibration Method   *172*
7.1.2   Procedure for Vision-Based Cable Tension Estimation   *173*
7.2   Implementation in the Hard Rock Stadium Renovation Project   *174*
7.2.1   Hard Rock Stadium   *175*
7.2.2   Test Description   *176*
7.2.3   Estimating and Validating Cable Force   *178*
7.3   Implementation in the Bronx-Whitestone Bridge Suspender Replacement Project   *184*
7.3.1   Bronx-Whitestone Bridge   *184*
7.3.2   Estimating Suspender Tension   *185*
7.4   Summary   *187*

8 Achievements, Challenges, and Opportunities *191*

8.1 Capabilities of Vision-Based Displacement Sensors: A Summary *191*

8.1.1 Artificial vs. Natural Targets *192*

8.1.2 Single-Point vs. Multipoint Measurements *192*

8.1.3 Pixel vs. Subpixel Resolution *193*

8.1.4 2D vs. 3D Measurements *194*

8.1.5 Real Time vs. Post Processing *194*

8.2 Sources of Error in Vision-Based Displacement Sensors *195*

8.2.1 Camera Motion *196*

8.2.2 Coordinate Conversion *197*

8.2.3 Hardware Limitations *198*

8.2.4 Environmental Sources *198*

8.3 Vision-Based Displacement Sensors for Structural Health Monitoring *199*

8.3.1 Dynamic Displacement Measurement *199*

8.3.2 Modal Property Identification *201*

8.3.3 Model Updating and Damage Detection *202*

8.3.4 Cable Force Estimation *203*

8.4 Other Civil and Structural Engineering Applications *204*

8.4.1 Automated Machine Visual Inspection *204*

8.4.2 Onsite Construction Tracking and Safety Monitoring *206*

8.4.3 Vehicle Load Estimation *206*

8.4.4 Other Applications *207*

8.5 Future Research Directions *208*

Appendix: Fundamentals of Digital Image Processing Using MATLAB *211*

A.1 Digital Image Representation *211*

A.2 Noise Removal *214*

A.3 Edge Detection *216*

A.4 Discrete Fourier Transform *217*

References *221*

Index *229*

List of Figures

1.1 Common displacement sensors: (a) LVDT; (b) laser vibrometer; (c) GPS. 4

1.2 Vision-based remote displacement sensor. 5

2.1 Commercially available video cameras: (a) CMOS image sensor with optical lens; (b) Camcorder; (c) DSLR camera; (d) PTZ security camera. 13

2.2 Vision-based multi-camera measurement system. 13

2.3 Time synchronization. 14

2.4 Procedure for 2D vision sensor implementation. 15

2.5 Defining a template subset in a source image. 17

2.6 Surface plot of the NCC and template coordinates in image 1. 17

2.7 Surface plot of the NCC and template coordinates in image 2. 17

2.8 Schematic of stereo camera calibration. 23

2.9 Scaling factor determination: (a) optical axis perpendicular to the object surface; (b) optical axis non-perpendicular to the object surface. 25

2.10 Error resulting from camera non-perpendicularity: (a) effect of optical axis tilt angle (f = 50 mm); (b) effect of the focal length of the lens ($\theta = 3°$). 27

2.11 Flowchart of the UCC implementation. 29

2.12 Orientation code ($N = 16$). 33

2.13 Matching results for images in ill conditions: (a) searching for a partially occluded toy, (b) highlighted CD jacket. 37

2.14 Bilinear interpolation for sub-pixel analysis. 38

2.15 Flowchart of vision sensor based on OCM. 38

2.16 User interface of the OCM-based displacement measurement software. 39

3.1 Shaking table test. 44

3.2 Comparison of sinusoidal displacements by the LVDT and vision sensor with an artificial target panel: (a) 1Hz sinusoidal signal; (b) 5Hz sinusoidal signal; (c) 10Hz sinusoidal signal; (d) 20Hz sinusoidal signal. 45

3.3 Comparison of earthquake displacements by the LVDT and vision sensor with a natural target. 45

3.4 Shaking table test of a three-story frame structure: (a) shaking table and frame structure; (b) vision sensor system. 47

3.5 Subpixel resolution evaluation using a UCC-based vision sensor: (a) resolution: ± 1.338 mm; (b) resolution: ± 0.669 mm; (c) resolution: ± 0.268 mm; (d) resolution: ± 0.067 mm.   49

3.6 Comparison of displacements by OCM (artificial target), UCC (artificial target) and LDS: (a) base displacement; (b) first-floor relative displacement; (c) second-floor relative displacement; (d) third-floor relative displacement.   50

3.7 Comparison of displacements by OCM (natural target), UCC (natural target), and LDS: (a) base displacement; (b) first-floor relative displacement; (c) second-floor relative displacement; (d) third-floor relative displacement.   52

3.8 Evaluation of robustness in unfavorable conditions.   53

3.9 Case 1 comparison: (a) displacements by OCM and UCC; (b) UCC cross-correlation function contour.   54

3.10 Case 2 comparison: (a) displacements by OCM and UCC; (b) UCC cross-correlation function contour.   55

3.11 Case 3 comparison: (a) displacements by OCM and UCC; (b) UCC cross-correlation function contour.   56

3.12 Case 4 comparison: (a) displacements by OCM and UCC; (b) UCC cross-correlation function contour.   57

3.13 A steel building frame model on a seismic shaking table.   58

3.14 Seismic shaking table setup.   59

3.15 Experimental results of the seismic shaking table test: (a) measured displacement by the vision sensor; (b) power spectral distribution.   60

3.16 Test setup for the simply supported beam.   61

3.17 Schematic of sensor placement.   61

3.18 Case of a non-perpendicular camera optical lens axis.   61

3.19 Images of a marker panel for different camera tilt angles: (a) 3°; (b) 5°; (c) 9°.   62

3.20 Comparison of displacement measurement at point 16: (a) camera tilt angle 3°; (b) camera tilt angle 5°; (c) camera tilt angle 9°.   62

3.21 Field test: (a) Streicker Bridge; (b) artificial target.   64

3.22 Randomly running pedestrians: displacement measurement by vision sensor: (a) measured displacement time history; (b) power spectral distribution.   64

3.23 Randomly running pedestrians: acceleration measurement: (a) measured acceleration time history; (b) power spectral distribution.   65

3.24 Jumping pedestrians: displacement measurement by vision sensor: (a) measured displacement time history; (b) power spectral distribution.   65

3.25 Jumping pedestrians: acceleration measurement: (a) measured acceleration time history; (b) power spectral distribution. 65

3.26 Field test on a highway bridge. 66

3.27 Experimental results of field tests on a highway bridge: (a) v = 3 km/h; (b) v = 50 km/h. 68

3.28 View of the two testbed bridges. 69

3.29 Field tests on the railway bridge: (a) setup of field tests; (b) schematic representation of the position between the camera and the bridge girders (c) remote measurement of bridge displacement under moving trainloads. 70

3.30 Target panel and existing features on the railway bridges: (a) HCB bridge; (b) steel bridge; (c) LED lights for night tests. 71

3.31 Test H1: comparison of displacements by three sensors (day). 72

3.32 Test H2: comparison of displacements by three sensors (day). 73

3.33 Test H3: comparison of displacements by three sensors (day). 73

3.34 Test H4: comparison of displacements by three sensors (day). 74

3.35 Test S1: comparison of displacements by two sensors (night). 74

3.36 Test S2: comparison of displacements by two sensors (night). 75

3.37 Test S3: comparison of displacements (night). 76

3.38 Test S4: comparison of displacements (night). 76

3.39 Schematic illustration of the displacement peak. 77

3.40 Errors between peak displacements of test H1–H4 of the HCB bridge. 78

3.41 Errors between peak displacements of the steel bridge in tests S1–S4. 78

3.42 Field test of the Vincent Thomas Bridge: (a) Vincent Thomas Bridge (Los Angeles, CA); (b) test setup. 81

3.43 Actual images captured by two cameras: (a) artificial target panel; (b) natural rivet pattern. 82

3.44 Displacement time histories: (a) measurement in the morning; (b) measurement in the evening. 83

3.45 Power spectral distribution: (a) measurement in the morning; (b) measurement in the evening. 83

3.46 Manhattan Bridge: (a) cross-section; (b) test setup. 84

3.47 Tracking target on the bridge: (a) one target; (b) simultaneous measurements of three targets. 85

3.48 Displacement measurement of one target. 85

3.49 Simultaneous displacement measurements of three targets. 86

3.50 Tracking targets on the bridge and the background building. 87

3.51 The camera motion and the mid-span vertical displacement of the bridge. 87

4.1   Typical modal testing and SHM systems using accelerometers:
(a) modal testing of a beam in the lab; (b) long-term SHM of
the Jamboree bridge.   90

4.2   Comparison of identified mode shapes of the frame structure.   93

4.3   Displacement measurements at points 2–31 by the vision sensor.   98

4.4   Comparison of displacement measurements (a) at point 9;
(b) at point 16.   99

4.5   Frequency results from (a) displacements at points 2–31 by the vision
sensor; (b) displacements at points 9 and 16 by LDS; (c) accelerations at
six points by accelerometers.   99

4.6   Comparison of mode shapes between the vision sensor and accelerometer:
(a) first mode shape; (b) second mode shape.   100

4.7   Stiffness optimization evolution using measurements taken by the vision
sensor (natural target).   103

4.8   Test setup: (a) beam; (b) camera; (c) schematics of intact and
damaged beams.   109

4.9   Displacement measurements at points 2–31.   110

4.10   Identified first two mode shapes of the intact and damaged beams:
(a) first mode shape; (b) second mode shape.   110

4.11   Damage indices of the damaged beam: (a) MSC damage index; (b) MMSC
damage index.   111

5.1   Railway bridge for model updating: (a) side view; (b) plan view;
(c) front view.   117

5.2   Freight train configuration.   117

5.3   Displacement history with train speed 8.05 km/h.   118

5.4   Schematic representation of the bridge-track-vehicle
interaction system.   119

5.5   Measured vs. simulated displacement using the initial FE model.   121

5.6   Sensitivity analysis procedure.   124

5.7   Objective functions w.r.t. normalized bridge equivalent stiffness R_{EI}:
(a) displacement; (b) acceleration.   125

5.8   Objective functions w.r.t. normalized bridge damping R_α:
(a) displacement; (b) acceleration.   126

5.9   Objective functions w.r.t. normalized rail bed stiffness $R_{k_{rb}}$:
(a) displacement; (b) acceleration.   127

5.10   Objective functions w.r.t. normalized rail bed damping $R_{c_{rb}}$:
(a) displacement; (b) acceleration.   128

5.11   Objective functions w.r.t. normalized train suspension stiffness R_{k_t}:
(a) displacement; (b) acceleration.   129

5.12   Objective functions w.r.t. normalized train suspension damping R_{c_t}:
(a) displacement; (b) acceleration.   130

5.13   Two-step FE model-updating procedure.   131

5.14   After Step 1: train speed update.   131

5.15   After Step 2: equivalent bridge stiffness update.   132

5.16   Bridge under a moving train.   132

5.17   Power spectral density (PSD) of measured displacement histories: (a) train speed = 7.93 km/h; (b) train speed = 36.80 km/h; (c) train speed = 70.22 km/h.   133

5.18   Computed displacement and acceleration time histories and their PSDs with train speed 7.93 km/h.   133

5.19   Computed displacement and acceleration time histories and their PSDs with train speed 36.80 km/h.   134

5.20   Computed displacement and acceleration time histories and their PSD with train speed 70.22 km/h.   134

5.21   Mid-span maximum displacements and accelerations w.r.t. different train speeds.   135

6.1   Schematics of the output-only simultaneous identification problem.   140

6.2   Output-only time-domain identification procedure.   144

6.3   Numerical example.   146

6.4   Effect of the number of sensors and noise level on the evolution of bridge stiffness identification: (a) two sensors; (b) three sensors; (c) seven sensors.   148

6.5   Identification errors for bridge stiffness.   149

6.6   Comparison of identified and reference impact forces considering noise: (a) noise-free; (b) 2% noise; (c) 5% noise; (d) 10% noise.   150

6.7   Comparison of predicted and reference/measured displacement responses at node 19: (a) noise-free; (b) 2% noise; (c) 5% noise; (d) 10% noise.   151

6.8   Effect of the initial stiffness value on the evolution of bridge stiffness identification: (a) 2% noise; (b) 5% noise; (c) 10% noise.   152

6.9   Effect of the damping estimate on the evolution of bridge stiffness identification.   153

6.10   Comparison of identified and reference impact forces considering damping error: (a) $\varsigma = 0.005$; (b) $\varsigma = 0.01$; (c) $\varsigma = 0.02$; (d) $\varsigma = 0.1$.   153

6.11   Impact test setup.   154

6.12   Measurement points.   154

6.13   Comparison of displacement measurements: (a) displacement at point 9; (b) displacement at point 16.   155

6.14   Beam stiffness identification from different initial stiffness values.   156

6.15   Identified and measured hammer impact forces.   157

6.16   Comparison of the predicted and measured beam displacement: (a) displacement at point 6; (b) displacement at point 23.   158

7.1   Outline of vision-based cable tension measurement.   173

7.2   Hard Rock Stadium.   175

7.3   Typical cable assembly.   178

7.4   Implementation of the computer vision sensor in Hard Rock Stadium: (a) tie down cables; (b) inclined cables.   178

7.5   Measured vibration and PSD function of TD_A cable at Quad A.   179

7.6   Measured vibration and PSD function of TD_B cable at Quad B.   179

7.7   Measured vibration and PSD function of TD_C cable at Quad C.   179

7.8   Measured vibration and PSD function of TD_D cable at Quad D.   179

7.9   Measured tension forces vs. reference forces for TD cables: (a) Quad A; (b) Quad B; (c) Quad C; (d) Quad D.   180

7.10   Measured vibration and PSD function of SLLB cable at Quad C.   181

7.11   Measured vibration and PSD function of EZUB cable at Quad C.   181

7.12   Measured vibration and PSD function of EZF cable at Quad C.   181

7.13   Measured vibration and PSD function of SLF cable at Quad C.   181

7.14   Measured tension forces for inclined cables using the vision sensor vs. reference values.   182

7.15   Bronx-Whitestone Bridge.   185

7.16   Suspender replacement locations.   185

7.17   Field suspender replacement: (a) jacking apparatus; (b) new tensioned suspender ropes.   186

7.18   Vision sensor setup for measuring suspender tension.   186

7.19   Measured vibration time histories and PSD amplitudes for suspender N61E: (a) suspender leg SL1; (b) suspender leg SL2; (c) suspender leg SL3; (d) suspender leg SL4.   188

8.1   Bridge inspection: (a) conventional visual inspection; (b) UAV inspection.   205

8.2   Examples of visible damage.   205

A.1   Examples of image types: (a) binary image; (b) grayscale image; (c) true-color image.   212

A.2   The 2-D Cartesian coordinate of an $M \times N$ grayscale image.   212

A.3   Noise removal example.   215

A.4   Edge detection example.   216

A.5   DFT of a grayscale image.   218

List of Tables

1.1   Comparison of sensors for measuring structural vibration.   6

2.1   Typical hardware components of vision sensor system.   12

3.1   Measurement errors of vision sensor in shaking table tests.   46

3.2   Different levels of subpixel resolution.   48

3.3   Measurement errors: NRMSE (%)   53

3.4   Test conditions of eight representative measurements.   71

3.5   Errors between peak displacements measured by different sensors.   79

4.1   Parameters of three-story frame structure.   92

4.2   Comparison of identified natural frequencies of the frame structure.   93

4.3   Parameters of the simply supported beam.   97

4.4   Comparison of identified natural frequencies of the beam structure.   100

4.5   Stiffness identification results ($\times 10^4 N/m$).   103

5.1   Design parameters of the bridge and track system.   120

5.2   Parameters of the freight train.   121

5.3   Dominant frequencies.   135

6.1   Parameters of numerical example.   146

6.2   Simulation cases.   147

7.1   Cable length, measured cable frequencies, and tension discrepancies w.r.t. reference values.   177

7.2   Cable geometric and material parameters.   187

7.3   Measured suspender rope frequencies and tension.   187

Series Preface

The Wiley-ASME Press Series in Mechanical Engineering brings together two established leaders in mechanical engineering publishing to deliver high-quality, peer-reviewed books covering topics of current interest to engineers and researchers worldwide.

The series publishes across the breadth of mechanical engineering, comprising research, design and development, and manufacturing. It includes monographs, references and course texts.

Prospective topics include emerging and advanced technologies in Engineering Design; Computer-Aided Design; Energy Conversion & Resources; Heat Transfer; Manufacturing & Processing; Systems & Devices; Renewable Energy; Robotics; and Biotechnology.

Preface

Over the past few decades, a significant number of studies have been conducted in the area of structural health monitoring (SHM), with the objective of detecting anomalies and quantitatively assessing structural integrity based on measurements using various types of sensors. Although these studies have produced SHM methods, frameworks, and algorithms that have been validated through numerical, laboratory, and field applications, their wide deployment in real-world engineering structures is limited by the prohibitive requirement of installing dense on-structure sensor networks and associated data-acquisition systems. To address these practical limitations, the research and industrial communities have been actively exploring new sensing technologies that can advance the current state-of-the-art in SHM.

Rapid advances in digital cameras and computer vision algorithms have made vision-based sensing a promising next-generation monitoring technology to complement conventional sensors. Significant advantages of the vision-based sensor include its low cost, ease of setup and operation, and flexibility to extract displacements at multiple points on the structure from a single video measurement. In the past 10 years, the authors have been fortunate to lead, participate in, and witness the development of computer vision-based sensing and its application to structural dynamics and SHM. In our activities, however, we have seen a gap between the significant potential offered by this emerging sensing technology and its practical applications. Many undergraduate and graduate students, researchers, and practicing engineers are interested in learning how this sensing technology works and what unique benefits it can offer.

This book is intended to provide a comprehensive introduction to vision-based sensing technology, based primarily on the authors' research. Fundamental knowledge, important issues, and practical techniques critical to the successful development of the vision-based sensor are presented and discussed in detail. A wide range of tests have been carried out in both laboratory and field environments to demonstrate its measurement accuracy and unique merits. The potential of the vision sensor as a fast and cost-effective tool for solving SHM problems is explored. In addition to SHM, novel and practical solutions to other engineering problems are presented, such as estimating cable tension forces using vision-based sensing. Finally, the book outlines the achievements and challenges of current vision-based sensing technologies, as well as open research challenges, to assist both the structural engineering and computer science research communities in setting an agenda for future research.

The goal of this book is to help encourage the application of the emerging vision-based sensing technology not only in scientific research but also in engineering practice, such as assessing the field condition of civil engineering structures and infrastructure systems. Although the book is conceived as an entity, its chapters are mostly self-contained and can serve as tutorials and reference works on their respective topics. The book may also serve as a textbook for graduate students, researchers, and practicing engineers; thus, much emphasis has been placed on making the computer vision algorithms, structural dynamics, and SHM applications easily accessible and understandable. To achieve this goal, we provide MATLAB code for most of the problems discussed in the book. In addition, readers working in structural dynamics and health monitoring will find this book hands-on and useful.

The authors would like to express their gratitude to the following individuals: Professors Shun'ichi Kaneko and Takayuki Tanaka at Hokkaido University, for inspiring the authors' work on computer vision more than a decade ago and for kindly providing the orientation code matching (OCM) MATLAB code included in Chapter 2; Dr. Yoshio Fukuda, former associate research scientist at Columbia University, for developing the OCM software package with the C++ language; Casey Megan Eckersley, PhD student of Columbia University, for her valuable assistance in editing the book; and last but not least, the authors' families for their strong support.

About the Companion Website

The companion website for this book is at

www.wiley.com/go/feng/structuralhealthmonitoring

The website includes: MATLAB CODES for chapters 2, 4, 6, 7 and appendix_matlab codes.

Scan this QR code to visit the companion website.

1

Introduction

1.1   Structural Health Monitoring: A Quick Review

Structures and civil infrastructure systems, including bridges, buildings, dams, and pipelines, are exposed to various external loads throughout their lifetimes. As they age and deteriorate, effective inspection, monitoring, and maintenance of these systems becomes increasingly important. However, conventional practice based on periodic human visual inspection is time-consuming, labor-intensive, subjective, and prone to human error. Nondestructive testing techniques have shown potential for detecting hidden damages, but the large size of the structural systems presents a significant challenge for conducting such localized tests. Over the past few decades, a significant number of studies have been conducted in the area of structural health monitoring (SHM), aiming at timely, objective detection of damage or anomalies and quantitative assessment of structural integrity and safety based on measurements by various on-structure sensors [1–4]. Most of the SHM techniques are based on structural dynamics, and the basic principle is that any structural damage or degradation would result in changes in structural dynamic responses as well as the corresponding modal characteristics. The SHM process is implemented in four key steps: data acquisition, system identification, condition assessment, and decision-making.

Dynamics-based SHM techniques can be categorized into frequency-domain and time-domain system identification methods. Carden and Fanning [5] presented an extensive literature review of frequency-domain SHM techniques based on changes in measured modal properties such as natural frequencies, mode shapes and their curvatures, modal flexibility and its derivatives, modal strain energy, frequency response functions, etc. Modal properties are obtained using various modal analysis techniques, e.g. the natural excitation technique, frequency

Computer Vision for Structural Dynamics and Health Monitoring, First Edition.
Dongming Feng and Maria Q. Feng.
© 2021 John Wiley & Sons Ltd.
This Work is a co-publication between John Wiley & Sons Ltd and ASME Press.
Companion website: www.wiley.com/go/feng/structuralhealthmonitoring

domain decomposition, stochastic subspace identification, the random decrement technique, blind source separation, and the autoregressive-moving-average model-fitting method. All of these methods have achieved satisfactory performance in numerical and experimental studies. For example, Kim and Stubbs [6] proposed a technique to locate and quantify cracks in beam-type structures based on a single damage indicator by using changes in natural frequencies. Lee et al. [7] presented a neural network–based method for element-level damage detection using mode shape differences between intact and damaged structures. Pandey et al. [8] proposed for the first time that mode shape curvature, which is the second derivative of the mode shape, is a sensitive indicator of damage. Feng et al. [9] developed the first neural network–based system identification framework for updating baseline structural models of two sensor-instrumented highway bridges.

Time-domain SHM techniques, rather than working with modal quantities, directly utilize measured structural response time histories to identify structural parameters. The identification in the time domain is often formulated as an optimization process, wherein the objective function is defined as the discrepancy between the measured and predicted responses. In the majority of existing studies, which are referred to as *input–output methods*, the known or measured excitation forces are a prerequisite for obtaining the predicted structural responses. However, it is highly difficult to measure excitation forces such as vehicle loads on bridges. Recently, there have been attempts to simultaneously identify both structural parameters and input forces from output-only identification formulations. For example, Rahneshin and Chierichetti [10] proposed an iterative algorithm – the extended load confluence algorithm – to predict dynamic structural responses in which limited or no information about the applied loads is available. Xu et al. [11] presented a weighted adaptive iterative least-squares estimation method to identify structural parameters and dynamic input loadings from incomplete measurements. Sun and Betti [12] demonstrated the effectiveness of a hybrid heuristic optimization strategy for simultaneous identification of structural parameters and input loads via three numerical examples. Feng et al. [13] proposed a numerical methodology to simultaneously identify bridge structural parameters and moving vehicle axle load histories from a limited number of acceleration measurements.

On the other hand, various filter-type algorithms for online system identification have been extensively studied in the literature, using either input–output or output-only time-domain data. Examples include the extended Kalman filter, unscented Kalman filter, particle filter, and H_∞ filter. For example, Chen and Feng [14] proposed a recursive Bayesian filtering approach to update structural parameters and their uncertainties in a probabilistic structural model. Soyoz and Feng [15] formulated an extended Kalman filter for instantaneous detection of seismic damage of bridges and validated its efficacy through large-scale seismic shaking-table tests. Although these online estimation algorithms have proved to be successful in many applications, they also present challenges. For example, the sensitivity of these methods to initial guess values affects the stability and convergence of estimated parameters to exact ones. In addition, parameter/damage identification methods based on heuristic algorithms – e.g. genetic algorithm, particle swarm optimization, artificial neural network, differential evolution, and artificial bee colony – have gained increasing attention due to their global optimization performance. However, validation of these methods is mostly limited to numerical or controlled laboratory examples rather than real-world structures.

For both frequency- and time-domain methods, vibration-based SHM strategies have proved effective in evaluating the global health state of structures and performing a rapid risk assessment. However, their wide deployment in realistic engineering structures is limited by the prohibitive requirement of installing dense on-structure sensor networks (primarily accelerometers) and associated data-acquisition systems. Contact-type wired sensors require time-consuming, labor-intensive installation and costly maintenance for successful long-term monitoring, which poses many economic and practical challenges. Although wireless sensor technology has addressed several limitations of wired sensors by eliminating cumbersome wiring, data acquisition remains challenging due to the complexity of data transmission, time synchronization, and power consumption, especially when hundreds of wireless sensors are mounted on a large-scale structure to measure dynamic responses. Moreover, one main bottleneck is that conventional on-structure sensors provide sparse, discrete point-wise measurements and thus low spatial-sensing resolutions, which limits the effectiveness of SHM on a large-scale structure. Although such a sensor network with a limited number of sensors may allow for the detection of changes in overall structural dynamics, it is often insufficient for identifying the location or assessing the extent of damage.

To address these practical limitations, the research and engineering practitioner communities have been actively exploring new sensor technologies that can advance the current state of SHM practice. This book introduces the emerging computer vision–based sensor technology.

1.2 Computer Vision Sensors for Structural Health Monitoring

While most SHM studies are based on the measurement of structural acceleration responses, displacement responses more directly reflect overall structural stiffness and thus offer the potential for improved accuracy in the assessment of structural conditions. As shown in Figure 1.1, sensors currently available for measuring structural displacements can be classified as contact types, such as the linear

(a) (b) (c)

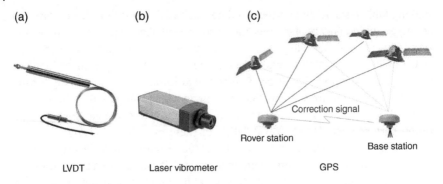

LVDT Laser vibrometer GPS

Figure 1.1 Common displacement sensors: (a) LVDT; (b) laser vibrometer; (c) GPS.

variable differential transformer (LVDT); and string potentiometer and noncontact types, such as GPS, laser vibrometers, and radar interferometry systems. These displacement sensors suffer from many limitations for field applications. For example, it is costly and highly difficult, if not impossible, to install an LVDT or a string potentiometer, which requires a stationary reference point; noncontact laser vibrometers are generally accurate but are costly and have a short measurement distance because of safety regulations; GPS sensors are easier to install, but the measurement accuracy is limited; and an interferometric radar system allows remote measurements with good resolution but requires reflecting surfaces mounted on the structure, which can be difficult to install and maintain.

Rapid advances in cameras and computer vision techniques have made vision-based sensing a promising alternative to conventional sensors for structural dynamic displacement measurement and health monitoring. As shown in Figure 1.2, a typical computer vision–based sensor system simply consists of one or more digital cameras and a computing unit such as a laptop or a tablet PC with measurement software installed. Video images of features on a structure, such as rivets and edges, are captured by the camera and streamed into the computer. By processing the digital video images using the measurement software, displacement time histories can be obtained at multiple locations simultaneously. The emerging vision-based sensor offers significant advantages over conventional contact-type and other noncontact-type displacement sensors, as summarized next [16]:

1) In contrast to a contact-type sensor (such as an LVDT or a string potentiometer), which requires time-consuming, costly installation on the structure and physical connections to a stationary reference point, a computer vision sensor requires no physical access to the structure, and the camera can be set up at a convenient remote location. This represents significant savings of both time and cost. For monitoring bridges, for example, no traffic control is required.

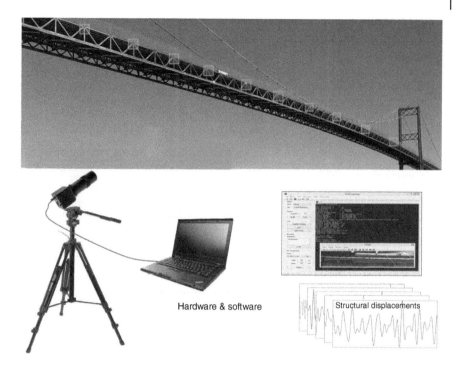

Figure 1.2 Vision-based remote displacement sensor.

In addition, each contact-type sensor measures one-dimensional (1D) displacement, but a single computer vision camera can measure two-dimensional (2D) displacements simultaneously.

2) Compared with a noncontact GPS, which requires installation on the structure (but not a stationary reference point), a vision-based sensor is far more accurate and less expensive. Depending on the cost, the GPS measurement error is typically in the range of 5–10 mm: more than an order of magnitude larger than that of a vision sensor.

3) Unlike a noncontact laser vibrometer, which must be placed very close to the measurement target due to the limited allowable laser power, a vision sensor can be placed hundreds of meters away (with the help of an appropriate zoom lens) and still achieve satisfactory measurement accuracy.

4) In contrast to conventional displacement sensors, almost all of which are point-wise sensors, a single vision sensor can simultaneously track structural displacements at multiple points. More importantly, one can easily alter the measurement points after video images are taken, offering unique flexibility for achieving better SHM results.

A comparison between commonly used vibration sensors and vision-based displacement sensors is summarized in Table 1.1.

Table 1.1 Comparison of sensors for measuring structural vibrations.

Sensors	Measure	Pros	Cons
Wired or wireless accelerometer	• Acceleration	• Suitable for continuous monitoring • Hardware easily available • Sensitive to high-frequency vibrations	• High cost of sensor system • High cost of installation and maintenance • Contact sensor • Single-point measurement • Additional mass on the structure may affect output
LVDT	• Displacement	• Hardware easily available	• Difficult and costly to install • Contact sensor • One-dimensional measurement • Single-point measurement
Laser vibrometer	• Velocity or displacement	• Noncontact • Accurate	• High cost of sensor system • Not suitable for continuous monitoring • Limited measurement distance
Computer vision sensor	• Displacement	• Noncontact, continuous monitoring • Low-cost industrial or consumer-grade video cameras • Two- or three-dimensional measurement • Multiple flexible measurement points on the visible object surface	• Accuracy affected by weather, light, and camera motion

About 10 years ago, the research community started to develop computer vision–based sensor technology for displacement measurement of large-size structures in controlled laboratory and challenging field environments. Modal analysis can be performed on the displacement data to extract natural frequencies and the mode shapes of a structure. Moreover, by analyzing the measured displacement time histories and modal analysis results, analytical models and parameters of the structure can be updated, damage detected, and structural integrity assessed. The adoption of vision sensors can significantly reduce the testing cost and time associated with conventional instrumentations. For example, Poozesh et al. [17] pointed out that testing a typical 50 m utility-scale wind turbine blade requires approximately 200 gages (costing $35 000–$50 000) and about three weeks to set up a conventional strain gauge system, while by contrast, a multicamera system could streamline the blade-testing process by eliminating the sensor instrumentation and reducing the setup time to two days.

It should be noted that computer vision sensing has been attracting attention and gaining popularity in two major areas of structural engineering: (i) vision-based sensors for displacement measurement and their SHM applications for modal/parameter identification, damage detection, force estimation, and model validation and updating; and (ii) visual monitoring of structural surface for defect detection and condition assessment, including the use of unmanned aerial vehicles (UAVs) and machine learning techniques. The emphasis of this book is on the former application.

1.3 Organization of the Book

The goal of this book is to encourage the application of the emerging computer vision–based sensing technology not only in scientific research but also in engineering practice such as field condition assessment of civil engineering structures and infrastructure systems. This book may serve as a textbook for graduate students, researchers, and practicing engineers. Thus much emphasis has been placed on making computer vision algorithms and their applications in structural dynamics and SHM easily accessible and understandable. To achieve this goal, throughout the book, MATLAB computer code is provided for most of the problems that are discussed. Even though the book is conceived as an entity, its chapters are mostly self-contained and can serve as tutorials and reference works on their respective topics.

Chapter 2 introduces fundamental facts about computer vision sensor systems and algorithms and software for measuring displacement time histories from video images. General principles are presented, including various template-matching techniques for tracking targets and coordinate-conversion methods for converting image pixel displacements to physical displacements. Vision sensor software packages are developed for real-time multipoint displacement measurement based on two representative template-matching techniques: upsampled cross-correlation (UCC) and orientation code matching (OCM).

Chapter 3 presents a wide range of tests conducted in both laboratory and field environments to evaluate the performance of the vision-based sensor system for dynamic displacement measurement. The accuracy of the measured displacement time histories is evaluated by comparing vision sensor results from tracking high-contrast artificial targets or low-contrast natural targets on the structural surface with those obtained with conventional reference sensors. The robustness of the vision sensor is examined against adverse environmental conditions such as dim light, background image disturbance, and partial template occlusion. The vision sensor system is also tested on outdoor in situ structures, including a pedestrian bridge, a highway bridge, two railway bridges, and two long-span suspension bridges. Dynamic displacements induced by various excitations are measured during the daytime and at night from different distances with and without artificial targets installed. These tests confirm the efficacy of the computer vision sensor system for measuring structural dynamic responses in outdoor environments.

Chapters 4–7 demonstrate the use of measured displacement data for SHM. **Chapter 4** compares modal analysis results based on displacement response data with those from conventional acceleration data. Furthermore, the identified modal parameters are used to update structural parameters such as the stiffness of a three-story frame structure and to detect damage in a beam structure.

Chapter 5 describes a model-updating approach for railway bridges, which is based on time-domain optimization of analytical models using in situ measurement of the bridge displacement time histories under trainloads. A finite element model of the bridge is developed, considering the train-track-bridge dynamic interaction. A sensitivity analysis investigates the intrinsic effects of parameters of the train, track, and bridge subsystems on the dynamic response of the bridge. The model-updating approach is applied to a short-span bridge to identify train parameters such as speed as well as bridge structural parameters such as stiffness. The computer vision–based model updating approach can be developed into an effective tool for long-term SHM of short-span railway bridges.

Chapter 6 explores a method for simultaneous identification of structural parameters and unknown excitation forces by using only displacement response (i.e. output-only), as in reality it is often highly difficult to measure excitation forces (i.e. input). Numerical analysis investigates the accuracy, convergence, and robustness of the identified results. Laboratory experiments on a beam structure accurately identified the hammer excitation forces as well as the beam stiffness from the beam displacement response measured by a single camera, validating this output-only method and demonstrating its potential for low-cost, long-term SHM.

Chapter 7 presents the application of the computer vision sensor for cost-effective estimation of tension forces in cables, the most important component in cable-supported bridges and roof structures. Compared with the existing vibration method based on acceleration measurements, which requires the installation of sensors on the cable, noncontact computer vision measurement of the cable vibration represents significant time and cost savings. This computer vision–based method is implemented in two engineering projects to estimate the cable forces of the cable-supported roof structure of the Hard Rock Stadium in Florida and the suspender forces of the Bronx-Whitestone Bridge in New York. Satisfactory agreement is found between cable forces measured by the vision-based sensor and conventional accelerometers.

Chapter 8 provides an overview of the achievements made thus far in computer vision sensor technology through a state-of-the-art literature review as well as a summary of this book. It also discusses challenges and opportunities, which the authors hope will inspire continued research on an extended adoption of computer vision technology for solving civil and structural engineering problems.

Appendix A further introduces the fundamentals of digital image processing using MATLAB, including digital image representation, noise removal, edge detection, and discrete Fourier transform.

2

Development of a Computer Vision Sensor for Structural Displacement Measurement

A computer vision sensor system measures displacements of an object by tracking pixel movements of selected targets on the object from video images captured by a digital camera. This chapter first describes typical hardware components for a computer vision sensor system and then introduces fundamental knowledge, algorithms, and computer vision software for measuring displacement time histories from video images. General principles are presented, including various template-matching techniques for tracking targets and coordinate conversion methods for determining calibration factors to convert image pixel displacements to physical displacements. Vision sensor software packages are developed for real-time multipoint displacement measurement based on two representative template-matching techniques: upsampled cross-correlation (UCC) and orientation code matching (OCM).

2.1 Vision Sensor System Hardware

A computer vision sensor system typically consists of a digital video camera, a zoom lens, and a computing unit. Table 2.1 shows a typical vision sensor system example using off-the-shelf hardware components, along with some important technical specifications. A digital camera is equipped with a charge-coupled device (CCD) or complementary metal-oxide semiconductor (CMOS) image sensor, which turns light into discrete signals. The *resolution* of a digital camera refers to the pixel count contained in an image, commonly stated in columns by rows. The *frame rate* is the number of images (frames) captured in a given time period, using frames per second (FPS).

Computer Vision for Structural Dynamics and Health Monitoring, First Edition.
Dongming Feng and Maria Q. Feng.
© 2021 John Wiley & Sons Ltd.
This Work is a co-publication between John Wiley & Sons Ltd and ASME Press.
Companion website: www.wiley.com/go/feng/structuralhealthmonitoring

Table 2.1 Typical hardware components of a vision sensor system.

Component	Model	Technical specifications
Video camera	Point Grey FL3-U3-13Y3M-C	Maximum resolution: 1280×1024 Frame rate: 150 FPS Chroma: Mono Sensor type: CMOS Pixel size: $4.8\,\mu m$
Optical lens	Kowa LMVZ990 IR	Focal length: 9–90 mm Maximum aperture: F1.8
Computing unit	Sony PCG-41216L	Intel Core i7-2620M CPU @ 2.70 GHz 8 GB RAM 250 HDD

In addtion to low-cost CCD/CMOS image sensors, other commercially available camcorders, digital single-lens reflex (DSLR) cameras, and pan, tilt, and zoom (PTZ) security cameras, as shown in Figure 2.1, can also be used in computer vision systems. As camera and computer technologies advance, the cost of these hardware components is expected to continue decreasing.

Figure 1.2 illustrates the concept of measuring structural displacement using a computer vision sensor system. The camera is fixed on a tripod (or any fixed base) and placed at a convenient remote location to take video of the structure. The computing unit acquires and processes the video images to extract displacements of the selected targets on the structure. If the software has real-time processing capabilities, the measured displacement time histories can be displayed on the computer screen in real time and automatically saved to the computer or transmitted to users in a remote location through a wired or wireless network. The video images can also be saved for post-processing, during which the measurement points can be altered. This flexibility of selecting any measurement points from a single measurement is a significant advantage of a computer vision sensor that cannot be achieved with conventional sensors installed at fixed locations.

When conducting multi-point measurements of a large structure, multiple synchronized cameras can be used, with each camera targeting a different section of the structure. Figure 2.2 depicts the configuration of a time-synchronized multi-point measurement system. The system consists of a wireless local area network (LAN) and multiple vision sensor subsystems controlled by one master computer.

(a) CMOS image sensor
with optical lens

(b) Camcorder

(c) DSLR camera

(d) PTZ security camera

Figure 2.1 Commercially available video cameras: (a) CMOS image sensor with optical lens; (b) Camcorder; (c) DSLR camera; (d) PTZ security camera. *Source:* Courtesy of Shutterstock.

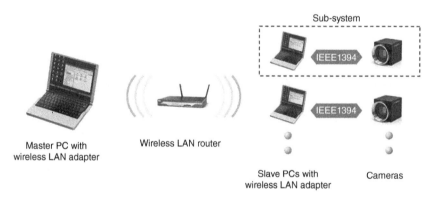

Sub-system

IEEE1394

IEEE1394

Master PC with
wireless LAN adapter

Wireless LAN router

Slave PCs with
wireless LAN adapter

Cameras

Figure 2.2 Vision-based multi-camera measurement system. *Source:* Reproduced with permission of John Wiley & Sons.

The subsystem slave computers are networked in one LAN via the wireless access point. In this network, the computers communicate with each other using TCP/IP. The displacement data measured by the subsystems is periodically sent to the master computer with a time stamp based on the subsystem computer.

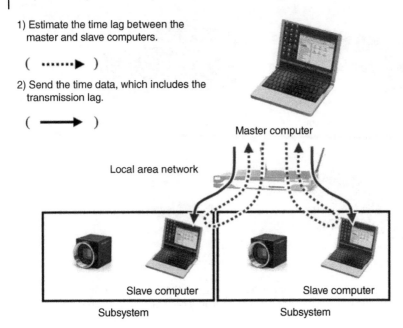

1) Estimate the time lag between the master and slave computers.

(·······▶)

2) Send the time data, which includes the transmission lag.

(——▶)

Master computer

Local area network

Slave computer   Slave computer

Subsystem   Subsystem

Figure 2.3 Time synchronization. *Source:* Reproduced with permission of John Wiley & Sons.

On the master computer, all the measured displacement data from the subsystems can be synchronized, displayed, and recorded.

To synchronize the recording times among all the subsystems, the master computer adjusts the local time of the subsystem slave computers by measuring the time delay of the wireless communications. Figure 2.3 shows how to synchronize the time clock of each computer. Basically, the internal clocks in all the subsystem slave computers are adjusted based on the internal clock in the master computer. First, the master computer sends each slave computer a dummy time that has the same data size as the actual time (dashed arrow in the figure). The slave computers immediately return the dummy time to the master computer. By measuring the time lag between the sending and receiving times, the master computer estimates the time cost to send the time data to the slave computers. Next, the master computer sends the time obtained from its internal time clock to each slave computer, and the slave computer considers the transmission lag, as shown in the solid arrow in the figure. Finally, the slave computers adjust their respective internal clocks according to the received time from the master computer.

2.2 Vision Sensor System Software: Template-Matching Techniques

Computer vision–based displacement sensors primarily utilize template matching – one of the most effective image-processing techniques – to track objects. Figure 2.4 shows the basic procedure for implementing two-dimensional (2D) displacement measurement. An initial area to be tracked is defined as a template in the first image of a sequence of video frames. The template can be located in the successive images using a template-matching technique. Template matching is a computationally intensive process. To reduce computation time, the search area can be confined to a predefined region of interest (ROI) near the template's location in the previous image. The displacement of the template is measured based on the pixel coordinates and converted into the physical displacement of the object using a scaling factor.

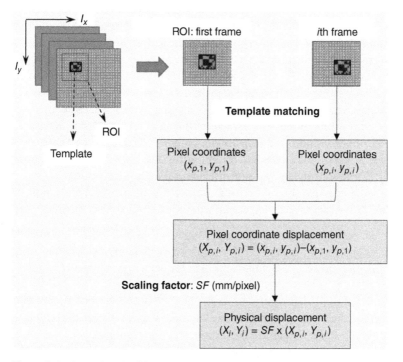

Figure 2.4 Procedure for 2D vision sensor implementation.

2.2.1   Area-Based Template Matching

Template matching techniques can be generally classified as *area-based* or *feature-based*. Area-based template matching methods, sometimes referred to as *correlation-like methods*, emphasize the matching step rather than the detection of salient objects. Classical area-based methods directly match image intensities using an exhaustive search strategy. To identify the matching area, the template image **T** is compared against the source image **I** by moving the template one pixel at a time (e.g. left to right, up to down). At each location, a metric is calculated to represent the similarity between the template image and the particular area of the source image. For each location (x, y) of **T** over **I**, the match metric is stored. The position of the template in the source image is determined by searching the peak position of the distribution of the match metric. The differences in the position of the template in video images yield the in-plane displacement vector, as illustrated in Figure 2.4.

The existing correlation criteria or similarity metrics/measures employed for vision sensors are often categorized into two groups [18]: (i) the cross-correlation (CC) criterion, which includes CC, normalized cross-correlation (NCC), and zero-normalized cross-correlation (ZNCC); and (ii) the sum of squared differences (SSD) correlation criterion, which includes SSD, normalized sum of squared differences (NSSD), and zero-normalized sum of squared differences (ZNSSD). Studies reveal that the ZNCC and ZNSSD correlation criteria offer the most robust noise-proof performance and are insensitive to the offset and linear scale in illumination lighting; the NCC and NSSD correlation criteria are insensitive to the linear scale in illumination lighting but are sensitive to offset of the lighting; and the CC and SSD correlation criteria are sensitive to all lighting fluctuations. All these methods have been successfully applied for structural displacement/deflection measurement. For example, vision-based displacement measurement systems were developed by Ye et al. [19] based on the NCC criterion, by Dworakowski et al. [20] based on the ZNCC criterion, and by Pan et al. [21] using the ZNSSD criterion.

The following example shows how to find the template displacement (in pixel coordinates) in two successive images using the conventional NCC method.

1) Define the template in the first image. As shown in Figure 2.5, a template image is defined as a subset of the first image in a sequence of video frames. In MATLAB, one can get these subregions using either a non-interactive or interactive script.

2) Conduct NCC to find the coordinates of a peak. Calculate the NCC, and display it as a surface plot. The peak of the CC matrix occurs where the template subset and the source image are best correlated. Thus template coordinates in the two successive source images can be found (Figures 2.6 and 2.7).

Figure 2.5 Defining a template subset in a source image.

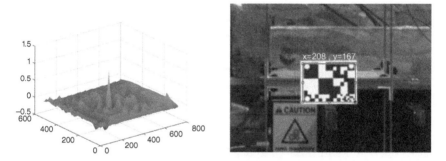

Figure 2.6 Surface plot of the NCC and template coordinates in image 1.

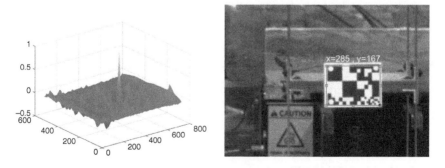

Figure 2.7 Surface plot of the NCC and template coordinates in image 2.

3) Calculate template pixel displacements in two source images. In this case, the translational template displacements are $(X_{p,2}, Y_{p,2}) = (x_{p,2}, y_{p,2}) - (x_{p,1}, y_{p,1}) = (285, 167) - (208, 167) = (77, 0)$.

The following are the MATLAB commands for this NCC template matching example. In order to demonstrate concepts and facilitate readers' self-learning and practice, this book lists most of the MATLAB code for the computer vision algorithms discussed. It should be noted that the reader will need access to licensed MATLAB software with the **Image Processing Toolbox** and the **Computer Vision System Toolbox**. Most of the MATLAB code presented in this book has been converted from the authors' code developed in other programming languages and is not optimized for MATLAB. Readers are encouraged to further improve the efficiency and performance of the code.

MATLAB Code – 2D Template Matching Using NCC

```
% Template matching using normalized 2-D cross-
correlation
% Syntax: C = normxcorr2(template, A)computes the
normalized cross-correlation of the matrices template
and A.
% The resulting matrix C contains the correlation
coefficients.

clc; clear; close all

%--------- Read two grayscale images for use with
"normxcorr2"
frame_1 = imread('videoframe_1.jpg');    % import the
1st video frame
frame_2 = imread('videoframe_2.jpg');    % import the
2nd video frame

%--------- Select Template T from 1st video frame
T=imcrop(frame_1);
imshow(T)
imwrite(T,'template.jpg')
close all

%--------- Perform cross-correlation, and display the
result as a surface
```

```
c1 = normxcorr2(T,frame_1);
c2 = normxcorr2(T,frame_2);
figure(1), surf(c1), shading flat
figure(2), surf(c2), shading flat

%--------- Find the peak in cross-correlation
[ypeak1, xpeak1] = find(c1==max(c1(:)));
[ypeak2, xpeak2] = find(c2==max(c2(:)));

%--------- Account for the padding that normxcorr2
adds
yoffSet1 = ypeak1-size(T,1);
xoffSet1 = xpeak1-size(T,2);
yoffSet2 = ypeak2-size(T,1);
xoffSet2 = xpeak2-size(T,2);

%--------- Display the matched area and position
figure(3)
imshow(frame_1);
imrect(gca, [xoffSet1+1, yoffSet1+1, size(T,2),
size(T,1)]);
hold on; plot(xoffSet1+size(T,2),yoffSet1+size(T,1),
'b*');
title('Template position in 1st image','FontName','Ari
al','fontsize', 17)
text(xoffSet1,yoffSet1-20,['x=',num2str(xoffSet1),' ,
y=',num2str(yoffSet1)],'FontName','Arial','fontsize',
18,'color','b')

% Show result
figure(4)
imshow(frame_2);
imrect(gca, [xoffSet2+1, yoffSet2+1, size(T,2),
size(T,1)]);
hold on; plot(xoffSet2+size(T,2),yoffSet2+size(T,1),'b*');
title('Template posizion in 2nd image','FontName','Ari
al','fontsize', 17)
text(xoffSet2,yoffSet2-20,['x=',num2str(xoffSet2),' ,
y=',num2str(yoffSet2)],'FontName','Arial','fontsize',
18,'color','b')
```

Comments:
1) normxcorr2 only works on grayscale images.
2) The scaling factor can be applied to convert the pixel coordinate displacement into a physical displacement.
3) To achieve subpixel precision with the NCC method, two approaches can be used. The first option is to resample the image intensity to a higher spatial resolution through interpolation. The second option is to interpolate the calculated CC surface to a higher spatial resolution; thus the CC peak can be located with a subpixel precision.

2.2.2 Feature-Based Template Matching

Feature-based template matching exhibits both geometric (i.e. translation, rotation, and scale) invariance and photometric (i.e. brightness and exposure) invariance. Feature detection, description, and matching are the three essential components in feature-based matching applications. The main steps often include: (i) detecting a set of distinctive key points and defining a region around each key point, (ii) computing and extracting local descriptors from normalized regions, and (iii) matching local descriptors.

Feature detection selects points of interest in an image that have unique features, such as corners (sharp image features) or blobs (smooth image features). The key to successful feature detection is to find features that remain locally invariant so that they can be detected even in the presence of illumination and scale changes, rotation, and occlusion. For example, the Harris features from accelerated segment test (FAST), and Shi and Tomasi methods can be adopted to detect corner features, and the speeded-up robust features (SURF), KAZE, and maximally stable extremal regions (MSER) methods for detecting blob features. On the other hand, feature extraction involves computing a compact vector representation of a local region centered on detected features. Descriptors, such as scale-invariant feature transform (SIFT) or SURF, rely on local gradient computations. Binary descriptors, such as binary robust invariant scalable keypoint (BRISK) or fast retina keypoint (FREAK), rely on pairs of local intensity differences, which are then encoded into a binary vector [16].

Once the local features and their descriptors have been extracted, the matching of feature points between two images can be performed by minimizing the Euclidean distance between the descriptors using nearest-neighbor matching algorithms. Two algorithms have been found to be most efficient for matching high-dimensional features: the randomized k-d forest and the fast library for approximate nearest neighbors (FLANN). It should be noted that these algorithms are not suitable for binary features (e.g. FREAK or BRISK), which can be compared using the Hamming distance calculated by performing a bitwise XOR operation followed by a bit count on the result [22]. In computer vision applications, to remove false matching points, statistically robust methods such as random sample consensus (RANSAC) can be used to filter outliers in matched feature sets while estimating the geometric transformation or fundamental matrix. For reference [23], Soh et al. compared several well-known techniques for feature selection and matching, i.e. the Kanade-Lucas-Tomasi (KLT) method, SURF with FLANN, SURF with brute force matching, and SIFT with RANSAC.

The following functions, provided by MATLAB Computer Vision System Toolbox, detect points of interest, extract feature descriptors, and enable automatic image registration.

MATLAB Code – Functions to Detect Interest Points and Extract Feature Descriptors

```
%------------ Detect Features ------------%
% Detect corners using Harris"CStephens algorithm and
return cornerPoints object
points = detectHarrisFeatures(I);
% Detect BRISK features and return BRISKPoints object
points = detectBRISKFeatures(I);
% Detect corners using FAST algorithm and return
cornerPoints object
points = detectFASTFeatures(I);
% Detect MSER features and return MSERRegions object
regions = detectMSERFeatures(I);
% Detect corners using minimum eigenvalue algorithm
and return cornerPoints object
points = detectMinEigenFeatures(I);
% Detect SURF features and return SURFPoints object
points = detectSURFFeatures(I);
% Detect KAZE features
points = detectKAZEFeatures(I);

%------------ Extract Features ------------%
% Extract histogram of oriented gradients (HOG) features
features = extractHOGFeatures(I);
% Extract local binary pattern (LBP) features
features = extractLBPFeatures(I);
% Extract interest point descriptors
[features,validPoints] = extractFeatures(I,points);
```

```
%------------ Match Features ------------%
% Find matching features
indexPairs = matchFeatures(features1,features2);
% Display corresponding feature points
showMatchedFeatures(I1,I2,matchedPoints1,matchedPoints2);

%------------ Image Registration ---------%
% Apply geometric transformation to image
B = imwarp(A,tform);
% Estimate geometric transform from matching point pairs
tform = estimateGeometricTransform(matchedPoints1,
matchedPoints2,transformType);
% Estimate motion between images or video frames
blkMatcher = vision.BlockMatcher;
% Find local maxima in matrices
LMaxFinder = vision.LocalMaximaFinder;
% Locate template in image
tMatcher = vision.TemplateMatcher;

%------------ Store Features ------------%
% Object for storing binary feature vectors
features= binaryFeatures(featureVectors);
% Object for storing BRISK interest points
points = BRISKPoints(Location);
% Object for storing KAZE interest points
points = KAZEPoints(location);
% Object for storing corner points
points = cornerPoints(location);
% Object for storing SURF interest points
points = SURFPoints(Location);
% Object for storing MSER regions
regions = MSERRegions(pixellist);
```

2.3 Coordinate Conversion and Scaling Factors

To measure the displacement of an object from captured video images, the relationship between image coordinates (in pixels) and physical/world coordinates (in millimeters or inches) must be established, based on which a scaling factor, i.e. the ratio between the physical and pixel displacements (in mm/pixel, for example), is determined. In general, the scaling factor can be determined based on

(i) the intrinsic parameters of the camera as well as the extrinsic parameters between the camera and the object, which can be obtained through camera calibration (the *camera calibration method*); or (ii) the known physical dimensions of the object surface and the corresponding image dimension in pixels (the *practical calibration method*).

2.3.1 Camera Calibration Method

Camera calibration is the process of estimating the camera parameters necessary to determine the scaling factors using images of a special calibration pattern. The parameters of interest include intrinsic camera parameters, distortion coefficients, and extrinsic camera parameters. Figure 2.8 shows a schematic of stereo calibration to estimate parameters for a pair of cameras, as well as the relative positions and orientations of the cameras [16]. After obtaining the camera parameters, the world coordinates of any image point can be reconstructed from its image coordinates.

The most frequently adopted pinhole model for camera calibration relates the three-dimensional (3D) world coordinates (X, Y, Z) of a calibration target point P

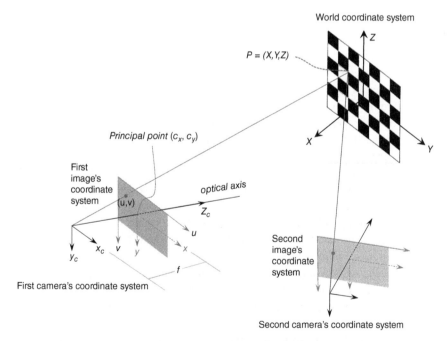

Figure 2.8 Schematic of stereo camera calibration. *Source:* Reproduced with permission of Elsevier.

with its corresponding location (u, v) in the image plane using a perspective transformation. The projection equation can be expressed as:

$$s\mathbf{m}' = \mathbf{A}[\mathbf{R}|\mathbf{T}]\mathbf{M}' \tag{2.1}$$

$$s\begin{bmatrix} u \\ v \\ 1 \end{bmatrix} = \begin{bmatrix} f_x & \gamma & c_x \\ 0 & f_y & c_y \\ 0 & 0 & 1 \end{bmatrix} \begin{bmatrix} r_{11} & r_{12} & r_{13} & t_1 \\ r_{21} & r_{22} & r_{23} & t_2 \\ r_{31} & r_{32} & r_{33} & t_3 \end{bmatrix} \begin{bmatrix} X \\ Y \\ Z \\ 1 \end{bmatrix} \tag{2.2}$$

where s is a scale factor, f_x and f_y are the horizontal and vertical focal lengths expressed in pixel, (c_x, c_y) is a principal point that is usually at the image center, and γ is a skew factor. The extrinsic parameters, $\mathbf{R}$ and $\mathbf{T}$, represent a rigid rotation and translation transformation from the 3D world coordinate system to the camera coordinate system. The intrinsic parameter $\mathbf{A}$ denotes a projective transformation from 3D camera coordinates into 2D image coordinates. The intrinsic parameter does not depend on the scene viewed. Therefore, once estimated, it can be reused as long as the focal length is fixed (if a zoom lens is used). Then, given the corresponding locations of the calibration control points in 3D world coordinates and 2D image coordinates, the unknown camera parameters can be estimated based on a nonlinear optimization, e.g. with the Levenberg–Marquardt algorithm. Wang et al. [24] summarized the total number of unknowns in both single- and stereo-camera calibrations.

Camera calibration can be conveniently conducted using the well-known OpenCV in the Open Source Computer Vision Library and the MATLAB camera calibration packages. To calibrate the camera, multiple images of a calibration pattern are needed from different angles. Popular calibration patterns include an asymmetric checkerboard (with one side containing an even number of squares, both black and white, and the other containing an odd number of squares) and circular control points. There are other calibration patterns as well. For example, Park et al. [25] used a T-shaped wand with multiple markers attached at predetermined positions to calibrate multiple cameras and set up the origin of the 3D displacement measurement. After the calibration, the wand is removed and not used during displacement measurement.

Coordinate conversion through camera calibration has been adopted for many vision-based displacement measurement systems. When a projection of 2D world coordinates to 2D image coordinates is considered, i.e. the objects are only subjected to in-plane motion, Wu et al. [26] proposed a simplified camera calibration method based on known world coordinates of at least four points. In the literature [27], the known dimensions from the edges and diagonals of the installed artificial target are used to establish the transformation between image and physical coordinates, with the assumption that out-of-plane motion is negligible.

2.3.2 Practical Calibration Method

The camera calibration method requires one-time access to the target object to install the calibration panel, which can be difficult in the field. For this reason, more practical calibration methods are desired. Described here is such a method developed by the authors [28].

As shown in Figure 2.9a, when the camera's optical axis is perpendicular to the object surface, all points on this surface have an equal depth of fields, meaning these points can be equally scaled down into the image plane. In this case, only a single scaling factor is needed. The scaling factor can be obtained using one of two methods:

$$SF = \frac{d_{known}}{I_{known}} \text{ or } SF = \frac{d_{known}}{d^i_{known}} d_{pixel} = \frac{D}{f} d_{pixel} \tag{2.3}$$

where d_{known} is the known physical length of the object surface, d^i_{known} and I_{known} are the corresponding physical length and pixel length, respectively, of the image plane with $d^i_{known} = I_{known} d_{pixel}$, d_{pixel} is the pixel size (e.g. in μm/pixel), D is the distance between the camera and the object, and f is the camera focal length.

However, the prerequisite of Eq. (2.3) is the perpendicularity of the camera's optical axis to the object surface such that all points on the object surface have an

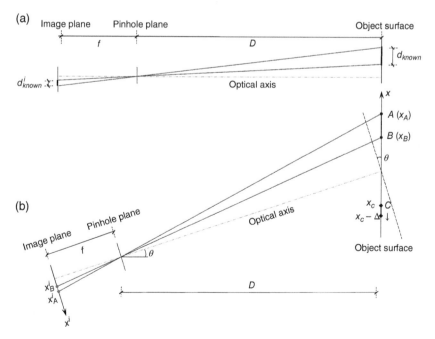

Figure 2.9 Scaling factor determination: (a) optical axis perpendicular to the object surface; (b) optical axis non-perpendicular to the object surface.

equal depth of field. Such a requirement would impose difficulties in the practical implementations because a small camera misalignment angle may be unnoticed during the experiment setup, especially when the object's distance from the camera is relatively large. Moreover, in outdoor field tests, there is sometimes no way to avoid tilting the camera optical axis by a small angle to track the measured object surface.

Figure 2.9b shows a schematic when the camera optical axis is tilted from the normal directions of the object surface by an angle θ. Assume line AB is a known dimension on the object. x_A and x_B are the coordinates of the two points, and I_A^i and I_B^i are the corresponding pixel coordinates at the image plane. The scaling factor can be estimated by:

$$SF_1 = \frac{x_A - x_B}{I_A^i - I_B^i} \tag{2.4}$$

From triangular geometry, x_A and x_B can be expressed as:

$$x_A = \frac{Dx_A^i}{f\cos^2\theta - x_A^i\cos\theta\sin\theta} \quad , \quad x_B = \frac{Dx_B^i}{f\cos^2\theta - x_B^i\cos\theta\sin\theta} \tag{2.5}$$

where $x_A^i = I_A^i d_{pixel}$ and $x_B^i = I_B^i d_{pixel}$ are the coordinates at the image plane. When θ is small $(\sin\theta\approx0)$ and $x_A^i \ll f$ and $x_B^i \ll f$, the scaling factor can be further estimated and simplified in terms of the intrinsic camera parameters and the extrinsic parameters between the camera and the object structure:

$$SF_2 = \frac{1}{I_A^i - I_B^i}\left(\frac{Dx_A^i}{f\cos^2\theta - x_A^i\cos\theta\sin\theta} - \frac{Dx_B^i}{f\cos^2\theta - x_B^i\cos\theta\sin\theta}\right)$$
$$\approx \frac{D}{f\cos^2\theta}d_{pixel} \tag{2.6}$$

For example, if point C in Figure 2.9b has a small translation Δ along the x-axis at the object surface, the *true displacement* is:

$$\Delta = x_C - \left(x_C - \Delta\right) = \frac{Dx_C^i}{f\cos^2\theta - x_C^i\cos\theta\sin\theta} - \frac{Dx_C^{i,\Delta}}{f\cos^2\theta - x_C^{i,\Delta}\cos\theta\sin\theta} \tag{2.7}$$

where $x_C^i = I_C^i d_{pixel}$ and $x_C^{i,\Delta} = I_C^{i,\Delta} d_{pixel}$ are the coordinates of point C before and after translation at the image plane.

From the scaling factors SF_1 in Eq. (2.4) and SF_2 in Eq. (2.6), the *measured displacement* can be estimated by $\tilde{\Delta}_1 = \left(I_C^i - I_C^{i,\Delta}\right)SF_1$ or $\tilde{\Delta}_2 = \left(I_C^i - I_C^{i,\Delta}\right)SF_2$. To quantify the error resulting from camera non-perpendicularity, numerical studies

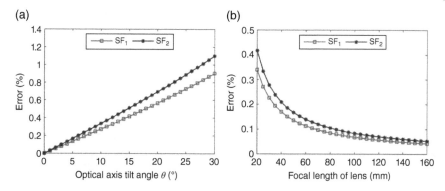

Figure 2.10 Error resulting from camera non-perpendicularity: (a) effect of optical axis tilt angle (f = 50 mm); (b) effect of the focal length of the lens (θ = 3°).

are conducted. The measurement errors from the two scaling factors can be defined as $Error = |\tilde{\Delta}_1 - \Delta| / \Delta \times 100\%$ and $Error = |\tilde{\Delta}_2 - \Delta| / \Delta \times 100\%$. In this study, the adopted parameters are as follows: camera with 640×512 pixel resolution, $d_{pixel} = 4.8\mu m$, $I_A^i = 200, I_B^i = 160$, and $D = 10$ m. Point C has a 1-pixel translation in the image plane from $I_C^i = 100$ to $I_C^{i,\Delta} = 99$.

The effects of the optical axis tilt angle and lens focal length are investigated numerically by considering a variable range, and the results are shown in Figure 2.10. It can be seen that the measurement error increases as the tilt angle increases, and the error is inversely related to the camera focal length. It could be concluded that in most practical applications, the measurement errors from small optical tilt angles are acceptable. Although this study is based on one-dimensional (x-axis) in-plane translation, the conclusions can be equally extended to 2D in-plane translation.

It is also found from the numerical study that for a fixed camera setup, the measurement error from scaling factor SF_1 in Eq. (2.4) decreases when the measurement point C gets closer to the known dimension AB. In particular, the error is minimized when the measurement point is located within the region of known dimensions. For scaling factor SF_2 in Eq. (2.6), errors further arise from uncertainties in the tilt angle estimation, camera distance measurement, and focal length readings from the adjustable-focal-length lens. In such cases, the fixed zoom lens and angle measurement system can be used to minimize the error.

For both the laboratory and field tests presented in this book, scaling factor SF_1 is adopted, which is obtained from Eq. (2.4) based on a known physical dimension on the object's surface (e.g. the size of artificial target panels or the size of nuts and rivets known from design drawings) and the corresponding image dimension in pixels.

2.4 Representative Template Matching Algorithms

Among the many template-matching algorithms, this book selects two representative algorithms – UCC and OCM – and demonstrates their implementation in computer vision software packages for measuring displacement. UCC uses cross-correlation by means of the Fourier transform in the image's spatial domain, while OCM is based on the gradient information of the image intensity in the form of orientation codes. Essentially, UCC is an area-based and OCM is a feature-based template-matching algorithm. They can also be distinguished in that UCC is based on image intensity and OCM on image gradient. These two representative algorithms are developed into software packages for real-time extraction of multipoint displacement time histories from video images.

2.4.1 Intensity-Based UCC Technique

The UCC subpixel template matching method was originally proposed by Guizar-Sicairos et al. [29]. Consider a pair of images $f(x, y)$ and $t(x, y)$ with identical dimensions $M \times N$, where $t(x, y)$ has a relative translation from the reference image $f(x, y)$. The cross-correlation between $f(x, y)$ and $t(x, y)$ by means of the Fourier transform can be defined as:

$$
\begin{aligned}
R_{FT}\left(x_0, y_0\right) &= \sum_{x,y} f\left(x, y\right) t^*\left(x - x_0, y - y_0\right) \\
&= \sum_{u,v} F\left(u, v\right) T^*\left(u, v\right) \exp\left[i2\pi\left(\frac{ux_0}{M} + \frac{uy_0}{N} \right) \right]
\end{aligned}
\tag{2.8}
$$

where the summations are taken over all image points (x, y); (x_0, y_0) is an amount of coordinate shift; the asterisk (*) denotes complex conjugation; and $F(u, v)$ and $T^*(u, v)$ represent the discrete Fourier transform (DFT) of their lowercase counterparts. For example,

$$
F\left(u, v\right) = \sum_{x,y} \frac{f\left(x, y\right)}{\sqrt{MN}} \exp\left[-i2\pi\left(\frac{ux}{M} + \frac{uy}{N} \right) \right]
\tag{2.9}
$$

From Eq. (2.8), an initial displacement estimation with pixel-level resolution can be easily acquired by locating the peak of R_{FT}. Subsequently, cross-correlation based on a time-efficient matrix-multiplication DFT is performed in a neighborhood around the initial peak to achieve subpixel resolution.

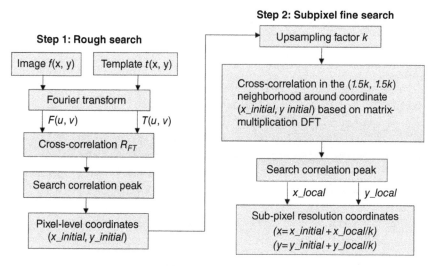

Figure 2.11 Flowchart of the UCC implementation.

Figure 2.11 shows the flowchart of the vision sensor based on subpixel UCC, described as follows:

1) *Pixel-level rough search.* Compute the cross-correlation between the image to register and the reference image using the Fourier transform. The initial displacement can be estimated from the correlation peak.
2) *Subpixel fine search.* Compute the cross-correlation in a 1.5 × 1.5 pixel neighborhood around the initial estimate by an upsampling factor of κ. Thus a subpixel resolution within $1/\kappa$ of a pixel is achieved by searching the peak in this (1.5κ, 1.5κ) neighborhood. For example, by setting $\kappa = 10$, a 0.1 subpixel accuracy can be achieved.

   In Step 2, instead of computing a zero-padded fast Fourier transform (FFT), a matrix-multiplication DFT operation is implemented by the product of three matrices with dimensions (1.5κ, N), (N, M), and (M, 1.5κ). The algorithm complexity for this upsampling subpixel search is $O(MN\kappa)$, while the complexity of FFT upsampled by zero-padding $F(u, v)T^*(u, v)$ is $O(MN\kappa[\log_2(\kappa M) + \kappa\log_2(\kappa N)])$. The substantial improvement dramatically reduces computational time and memory requirements without sacrificing accuracy, making possible real-time displacement measurements.

The following is the MATLAB script for using the UCC template matching technique to obtain displacement time histories from recorded video files.

MATLAB Code – Displacement Time History Measurement Using UCC

```
clc; clear; close all
tic
%-------- Import a AVI video file
path='C:\Bridge vibration.avi';
obj = VideoReader(path); % Construct "obj" to read
                                video data from file
N = obj.NumberOfFrames;   % Total number of frames in
                                the video stream
h = obj.Height;           % Height of the video frame
                                in pixels
w = obj.Width;            % Width of the video frame in
                                pixels
Fs = obj.FrameRate;       % Frame rate of the video in
                                frames per second
dt=1/Fs;
t=0:dt:(N-1)*dt;          % Generate time instants


%-------- Define Template subset in the 1st video
frame
% Convert RGB image (video frame 1) to grayscale and
double precision
frame1=im2double(rgb2gray(read(obj, 1)));
% Define template T in frame 1 using the interactive
Crop Image tool
T=imcrop(frame1);
[mt,nt]=size(T);
close all


%-------- Obtain the Scaling Factor
imshow(frame1)
% Identify 2 points and return their x- and y-coordinates
[x, y]=ginput(2);
% Known pixel length at the image plane
I_known=max(abs(x(2)-x(1)),abs(y(2)-y(1)));
% Input known physical length corresponding to the two
points
D_known= input('Input the known physical distance
(Units:mm): ');
```

```
SF=D_known/I_known;      % Scaling factor, units: mm/
pixel

%-------- Define region of interest (ROI)
px=round(h/20);
py=round(w/20);
% Define ROI near template's location in video frame 1
ROI_frame1=frame1(round(y(1)-
px):round(y(2)+px),round(x(1)-py):round(x(2)+py),:);
[mr,nr]=size(ROI_frame1);
T(mt+1:mr,:)=0;
T(:,nt+1:nr)=0;

%-------- Conduct UCC to obtain structural
displacement
buf2ft-fft2(T);         % Fourier transform of template
usfac=50;               % Upsampling factor (integer).

V=zeros(N,2);
Disp_P=zeros(N,2);
Disp_S=zeros(N,2);
for i = 1:N
% Convert RGB image (video frame i) to grayscale and
double precision
framei=im2double(rgb2gray(read(obj, i)));
% Define ROI near template's location in video frame i
ROI_framei=framei(round(y(1)-px):round(y(2)+px),
round(x(1)-py):round(x(2)+py),:);
buf1ft=fft2(ROI_framei);   % Fourier transform of
source image

% Conduct subpixel template matching by UCC
[output, Greg] = dftregistration(buf1ft,buf2ft,usfac);
% output(3) and (4) are row and column pixel shifts
between template and source images
V(i,:)=[output(3),output(4)];
% Compute Template pixel coordinate displacement w.r.t
template location in video frame 1
```

```
Disp_P(i,:)=V(i,:)-V(1,:);
% Convert the pixel coordinate displacement into
structural displacement
Disp_S(i,:)=Disp_P(i,:)*SF;
end
toc
save Displacement.mat Disp_S

figure(1)
subplot(2,1,1)
plot(t,Disp_S(:,1),'b')
xlabel('Time (s)', 'FontName','Arial','fontsize', 12)
ylabel('Displacement(mm)',
'FontName','Arial','fontsize', 12)
title('Displacement in X direction','FontName','Arial'
,'fontsize', 12)
subplot(2,1,2)
plot(t,Disp_S(:,2),'r')
xlabel('Time (s)', 'FontName','Arial','fontsize', 12)
ylabel('Displacement(mm)',
'FontName','Arial','fontsize', 12)
title('Displacement in Y direction','FontName','Arial'
,'fontsize', 12)
set(gcf,'Position',[200 200 1000 600]);
```

Comment:
1) Images will be registered within 1/usfac of a pixel. For example, usfac = 50 means the images will be registered within 1/50 of a pixel.
2) function [output Greg] = dftregistration(buf1ft,buf2ft,usfac) is the efficient subpixel image registration by UCC. This code was written by Manuel Guizar-Sicairos et al. and can be downloaded from https://www.mathworks.com/matlabcentral/fileexchange/18401-efficient-subpixel-image-registration-by-cross-correlation.

   This code gives the same precision as the FFT UCC in a small fraction of the computation time and with reduced memory requirements. It obtains an initial estimate of the cross-correlation peak by an FFT and then refines the shift estimation by upsampling the DFT in only a small neighborhood of that estimate by means of a matrix-multiply DFT. With this procedure, all the image points are used to compute the UCC in a very small neighborhood around its peak.

2.4.2   Gradient-Based Robust OCM Technique

The OCM method was originally proposed by Ullah and Kaneko [30] for robust template matching and further developed by the authors for displacement measurement [31]. In the OCM scheme, orientation code representations for an object image and the template are constructed from the corresponding gray images such that each pixel represents an orientation code that is obtained by quantizing the orientation angle at the corresponding pixel position in the gray image. The orientation angle represents the steepest ascent orientation evaluated from the pixel neighborhoods and measured with respect to the horizontal axis. The orientation codes thus obtained are a function of the texture and shape of the object and hence essentially invariant to object translation and the effects of shading, background, and illumination variations [32, 33].

Let an image be represented by $I(x, y)$ and the horizontal and vertical derivatives be $\nabla I_x = \partial I/\partial x$ and $\nabla I_y = \partial I/\partial y$, respectively. For the discrete version of the image, the orientation angle $\theta_{i,j}$ is computed by using the function $\theta_{i,j} = \tan^{-1}(\nabla I_y/\nabla I_x)$. Since the numerical value of the $\tan^{-1}$ function is confined to $[-\pi/2, \pi/2]$, the actual orientation is determined after checking signs of the derivatives, so the range of $\theta_{i,j}$ is $[0, 2\pi]$. The orientation code, as depicted in Figure 2.12, can be obtained by quantizing $\theta_{i,j}$ into N levels with a constant width $\Delta\theta$. An appropriate value of the width needs to be chosen to precisely detect the template instances in the scene. This issue should be considered in relation to the inherent amount of information and possible spatial resolution.

Figure 2.12   Orientation code (N = 16). *Source:* Reproduced with permission of John Wiley & Sons.

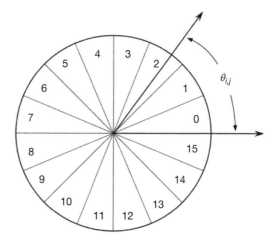

The orientation code is defined as follows using Gaussian notation:

$$c_{i,j} = \begin{cases} \left[\dfrac{\theta_{i,j}}{\Delta_\theta} \right], & \textit{if } |\nabla I_x| + |\nabla I_y| \geq \Gamma \\ N = \dfrac{2\pi}{\Delta_\theta}, & \textit{otherwise} \end{cases} \tag{2.10}$$

where Γ is a pre-specified threshold level for ignoring low-contrast pixels. The purpose of using Γ is to prevent uniform or semi-uniform regions from influencing the error evaluation, because pixels with low-contrast neighborhoods are more sensitive to noise. Using a value of Γ that is too large can cause the suppression of information in low-contrast images (containing strongly shaded or illuminated objects). Γ is set to 10 in this study. An example of the orientation codes is shown in Figure 2.14, corresponding to a quantization width of $\Delta\theta = \pi/8$. When the contrast is lower than the threshold Γ, then $N = 2\pi/\Delta\theta$ is assigned to the orientation code $c_{i,j}$.

A dissimilarity that is measured based on the definition of the orientation codes is designed to evaluate the difference between any two images of the same size. The best match between orientation code images of template **T** and any source image **I** from the same scene is found by minimizing the dissimilarity measure in the form of a summation of the error function, as follows:

$$D = \frac{1}{M} \sum_{I_{m,n}} d\left(O_{I_{m,n}}(i,j), O_T(i,j) \right) \tag{2.11}$$

where $O_{I_{m,n}}$ and O_T are the orientation code images of the sub-image and the template, respectively, M is the size of the template, (m, n) shows the position of the sub-image in the scene, and $d(\cdot)$ is the error function based on an absolute difference criterion:

$$d(a,b) = \begin{cases} \min\{|a-b|, N - |a-b|\}, & \textit{if } a \neq N \cap b \neq N \\ \dfrac{N}{4}, & \textit{if } a \neq N \cap b = N \\ 0, & \textit{if } a = N \end{cases} \tag{2.12}$$

When a comparison is performed between a pixel whose orientation code is evaluated by the $\tan^{-1}$ function and one whose code is set to N due to a low-contrast neighborhood, the error cannot be computed by finding the difference. To avoid such an inconsistent comparison, a reasonable value needs to be assigned to the error function corresponding to the pixels. The assigned value should not bias the dissimilarity evaluation for the sub-image. As an example, the value $N/4$

is assigned to the error function. This is the error value that is expected when these two pixels have no relation to each other. A large value of N is helpful for discriminating such an incompatible comparison.

Since the orientation code is cyclic, the absolute difference is not used directly to compute the error function; rather, the minimum distance between the two codes is determined. For example, in Figure 2.14, code 0 is only a unit away from code 15, whereas their simple difference yields 15. Such an evaluation stabilizes the match against minor pose variations of the object, regardless of the direction of the movement. As a consequence of this cyclic property of orientation codes, the maximum distance between any two codes is never more than $N/2$.

Finally, the similarity ratio S is derived as follows:

$$h = \frac{D}{N/2} \quad \left(0 \le h \le 1\right) \tag{2.13}$$

$$S = 1 - h \quad \left(0 \le S \le 1\right) \tag{2.14}$$

where h is the discrimination ratio obtained by dividing the average absolute difference D by the maximum of the absolute difference $N/2$. S is the similarity ratio.

The MATLAB commands listed next demonstrate how to find the template from the source image (defined in Figure 2.5) using the OCM technique.

MATLAB Code – Template Matching Using OCM

```
clc;clear;close all;
global del_theta N
%------------ Import source and template images
Itarg=imread('videoframe_1.jpg'); % load target image
Itemp=imread('template.jpg');      % load template image
[m1,n1]=size(Itarg);
[m2,n2]=size(Itemp);

%------------ Convert to orientation code image
del_theta=pi/8;
N=2*pi/del_theta;
OCtarg=orientation(m1,n1,Itarg);
OCtemp=orientation(m2,n2,Itemp);

for i1=1:m1-m2+1
for j1=1:n1-n2+1
        OCTARG=OCtarg(i1:i1+m2-1,j1:j1+n2-1);
        D=0;
```

```
% Calculate the dissimilarity pixel by pixel.
for i2=1:m2
for j2=1:n2
            d(i2,j2)=0;
if OCTARG(i2,j2)<N && OCtemp(i2,j2)<N
            d(i2,j2)=min(N-abs(OCTARG(i2,j2)-
OCtemp(i2,j2)),abs(OCTARG(i2,j2)-OCtemp(i2,j2)));
else if OCTARG(i2,j2)==N && OCtemp(i2,j2)==N
                d(i2,j2)=0;
else d(i2,j2)=N/4;
end
end
            D=D+d(i2,j2);
end
end
    S(i1,j1)=1-D/(8*(m2*n2));        % Get the similarity
end
end
%------------ Get the position of maximum similarity
[ypeak, xpeak] = find(S==max(S(:)));
figure
imshow(Itarg);
imrect(gca, [xpeak, ypeak, size(Itemp,2), size(Itemp,1)]);
                                % Mark the matched image.

function code = orientation( m,n,I )
global del_theta N
%------------ Sobel operator
sobelx=[-1 0 1;-2 0 2;-1 0 1];
sobely=[-1 -2 -1;0 0 0;1 2 1];
%------------ Gradient computation
Gx=conv2(I,sobelx,'same');
Gy=conv2(I,sobely,'same');
%------------ Threshold for ignoring the low-contrast
pixels
Gamma=10;
%------------ Orientation coding
for i=1:m
for j=1:n
        thro=abs(Gx(i,j))+abs(Gy(i,j));
```

```
if thro>Gamma
        thta(i,j)=atan2(Gy(i,j),Gx(i,j))+pi;
        code(i,j)=floor(thta(i,j)/del_theta);
else
        code(i,j)=N;
end
end
end
end
```

To evaluate the robustness of the OCM technique for object searching and template matching in less-than-ideal conditions, test images are prepared, as shown in Figure 2.13, where the templates (a bear toy and a small part of a CD jacket, shown as inserts in the upper-right corner) undergo illumination variations due to background occlusion by a nearby box or appear highlighted due to surface reflection. Test results show that the OCM technique can successfully detect the correct position of the target template, whereas conventional intensity-based template matching techniques such as NCC and SSD fail to do so [33].

To further improve the image resolution, the OCM algorithm is implemented by interpolating the obtained orientation angle with bilinear interpolation. As illustrated in Figure 2.14, the interpolated orientation angle θ is obtained as follows [32]:

$$\theta = \beta\left(\alpha\theta_{11} + (1-\alpha)\theta_{10}\right) + (1-\beta)\left(\alpha\theta_{01} + (1-\alpha)\theta_{00}\right) \tag{2.15}$$

(a)   (b)

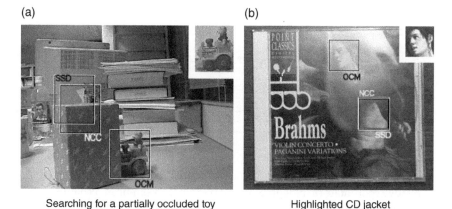

Searching for a partially occluded toy   Highlighted CD jacket

Figure 2.13 Matching results for images in ill conditions: (a) searching for a partially occluded toy, (b) highlighted CD jacket.

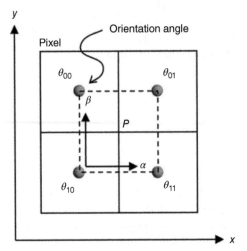

Figure 2.14 Bilinear interpolation for sub-pixel analysis. *Source:* Reproduced with permission of John Wiley & Sons.

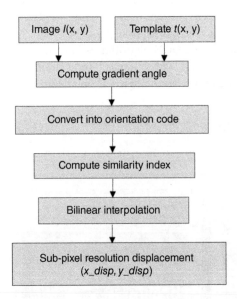

Figure 2.15 Flowchart of vision sensor based on OCM. *Source:* Reproduced with permission of John Wiley & Sons.

where θ_{00}, θ_{01}, θ_{10}, and θ_{11} are the orientation angles surrounding the grid point P. The relative coordinate (α, β) represents the position in a sub-pixel resolution. The range of each axis is $[0, 1]$.

Figure 2.15 presents the vision sensor procedures based on the sub-pixel OCM scheme.

2.4.3 Vision Sensor Software Package and Operation

The UCC and OCM algorithms described have been developed into software packages, which are used for the various laboratory and field evaluation experiments presented in this book. The programming environment is Visual Studio 2010, using C++ (which has been converted into the MATLAB code in this book to demonstrate the concepts and facilitate readers' self-learning). The FlyCapture software development kit (SDK) by Point Grey Research is used to capture video images from Point Grey USB 3.0 cameras using the same application programming interface (API) under a 32- or 64-bit Windows 7/8 operating system. The captured images are then processed frame by frame by the template matching algorithm and displayed on the screen using the DirectShow library. Efforts have been made to reduce the image processing time to realize real-time online measurements.

A user-friendly interface is built into the software package, as shown in Figure 2.16. The measured displacement history is both shown on the screen in real time and saved to the computer. The online measurement avoids time-consuming, memory-intensive storage of huge video files. However, a trade-off is necessary among measurement points, video resolution, maximum frame rate, and template sizes. The software can also save the video files for post-processing, which offers the flexibility to extract structural displacements at any desired point from a single recording.

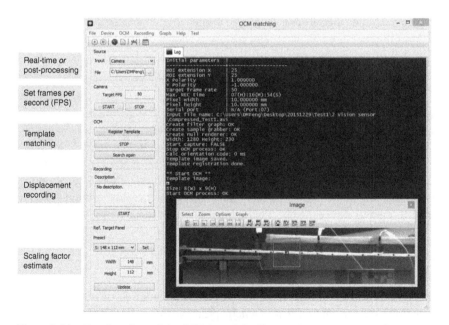

Figure 2.16 User interface of the OCM-based displacement measurement software.

The operation of the computer vision system for measuring displacements of a structure in real time is as follows [16]:

1) *Vision sensor system setup.* Fix the camera with a lens on a tripod (or any fixed base), and place it at a convenient location. Connect the camera to a computer that has the real-time displacement measurement software installed. It is noteworthy that setting up the vision sensor, including focusing the lens on the selected targets of the structure, takes only a few minutes.

2) *Scaling factor determination.* According to the practical calibration method described in Section 2.3.2, the scaling factor is estimated from a known physical dimension on the structural surface and its corresponding image dimensions in pixels. For example, if a 100 mm structural length is represented with 500 pixels, the scaling factor is 0.2 mm/pixel. When a structural dimension is not available, a reference target panel with a known dimension can be mounted on the structural surface for the calibration purpose.

3) *Single- or multiple-target/template registration.* Any natural marker on the structural surface can be registered as a tracking target, as long as it has pattern contrast compared with the surrounding background. Example natural markers include surface texture, bolts, and rivets.

4) *Template matching to extract displacement.* By clicking Start on the Recording module of the software, shown in Figure 2.16, video images captured by the camera are streamed into the computer, from which the UCC or OCM template matching algorithm is employed to track the targets registered in Step 3. Thus the measured displacement history is shown on the screen in real time and saved to the computer.

As noted in Step 4, it would be highly time-consuming (and would thus make real-time measurement impossible) if the target had to be searched for within the entire image of each video frame. To reduce the search time, the searching area can be confined to a predefined ROI near the template's location in the previous image. The new target ROI must be able to cover the target's potential new position in the next frame. Otherwise, mismatching will be induced. Also note that geometrical distortion due to lens optics, especially when a short focal length is used, can occur in the video images. In this case, a camera calibration process can be performed to reduce the effect of lens distortion [34].

2.5 Summary

This chapter introduced the fundamentals of a computer vision sensor system, including hardware and software. General principles were described by reviewing template matching techniques for tracking targets on a moving object and coordinate conversion methods for determining calibration factors to convert image pixel displacements to physical displacements. This chapter further presented the vision sensor software packages developed by the authors based on two representative template matching techniques: the image intensity-based subpixel UCC and image gradient-based robust OCM algorithms. The general procedure for operating the vision sensor system to measure structural displacement was discussed. Related MATLAB code was included to demonstrate the concept and facilitate readers' self-learning.

3

Performance Evaluation Through Laboratory and Field Tests

This chapter describes a wide range of experiments conducted in both controlled laboratory and challenging field environments to evaluate the performance of the computer vision sensor system for measuring dynamic displacement. Laboratory shaking table tests are carried out to compare the accuracy of displacements measured by vision sensors that track high-contrast artificial targets and low-contrast natural targets, respectively, on the structural surface as well as by reference sensors. The robustness of the vision sensor is examined in unfavorable environmental conditions such as dim light, background image disturbance, and partial template occlusion.

In outdoor environments, field tests are carried out on a pedestrian bridge on Princeton University campus, a highway bridge in Korea, two railway bridges in Pueblo, Colorado, and two long-span suspension bridges – the Vincent Thomas Bridge in San Pedro, California, and the Manhattan Bridge in New York City. The performance of the computer vision sensor system is evaluated in terms of its long-distance, multipoint measurement accuracy in the presence of various sources of outdoor environmental noise and lighting conditions.

3.1 Seismic Shaking Table Test

The accuracy of the vision-based sensor system in measuring dynamic displacements is evaluated through a shaking table test in the Carleton Laboratory at Columbia University. The shaking table is driven in the horizontal direction by two types of dynamic signals: sinusoidal signals of 1 through 20 Hz, and the 1941 El Centro Earthquake ground motion. As shown in Figure 3.1, a predesigned

Computer Vision for Structural Dynamics and Health Monitoring, First Edition.
Dongming Feng and Maria Q. Feng.
© 2021 John Wiley & Sons Ltd.
This Work is a co-publication between John Wiley & Sons Ltd and ASME Press.
Companion website: www.wiley.com/go/feng/structuralhealthmonitoring

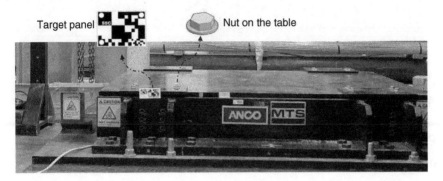

Figure 3.1 Shaking table test.

black and white artificial target panel (48 mm × 34 mm) is fixed on the shaking table and compared with a low-contrast target – the existing nuts on the table side surface.

The CMOS camera described in Table 2.1 is used. The camera equipped with a 90 mm lens is placed at a stationary position, 3.6 m from the shaking table. The camera captures the video images of the targets at a sampling rate of 150 fps. Each frame of the images is sent to the laptop; and using the developed software, the displacement of the shaking table is extracted in real time. The scaling factor for the vision sensor is determined to be 0.432 mm/pixel according to the practical method in Section 2.3.2 of Chapter 2. At the same time, the displacement is also measured by a high-accuracy, high-precision linear variable differential transformer (LVDT; model 0222-0000 by Trans-Tek Inc.), which is installed between the shaking table and a stationary reference point.

The displacement time histories measured by the computer vision sensor targeting the artificial and natural targets agree well with those measured by the reference LVDT. Figure 3.2 plots the displacement time histories measured by the orientation code matching (OCM) based vision sensor that targets the artificial target panel and by the LVDT under sinusoidal excitations with different frequencies: 1, 5, 10, and 20 Hz. Note that the amplitude of the sinusoidal displacement significantly decreases from 10 to 1 mm as the shaking frequency increases. Despite the small amplitudes at high frequencies, the displacements measured by the vision-based sensor agree well with those measured by the highly accurate and precise LVDT.

Figure 3.3 shows a comparison between the displacements measured by the OCM-based vision sensor and the LVDT under the seismic excitation. Instead of the artificial target panel, the computer vision sensor tracks the natural target: the existing nuts on the table, as shown in Figure 3.1. Excellent agreement is observed between the measurements by the LVDT and the vision sensor without using the high-contrast target panel.

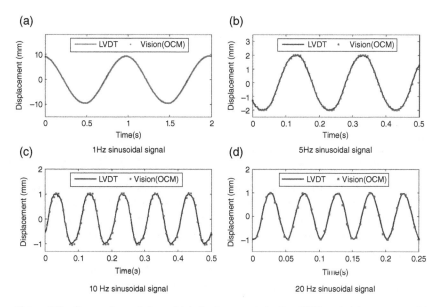

Figure 3.2 Comparison of sinusoidal displacements by the LVDT and vision sensor with an artificial target panel: (a) 1Hz sinusoidal signal; (b) 5Hz sinusoidal signal; (c) 10Hz sinusoidal signal; (d) 20Hz sinusoidal signal.

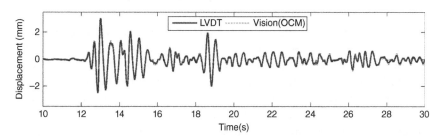

Figure 3.3 Comparison of earthquake displacements by the LVDT and vision sensor with a natural target.

To quantify the accuracy of the vision sensor, error analysis is performed using the normalized root mean squared error (NRMSE)

$$\text{NRMSE} = \frac{\sqrt{\dfrac{1}{n}\sum_{i=1}^{n}\left(x_i - y_i\right)^2}}{y_{\max} - y_{\min}} \times 100\% \tag{3.1}$$

where n = number of measurement data; x_i and y_i = ith displacement data at time t_i, measured by the vision sensor and the LVDT, respectively; and $y_{\max} = \max(y_i)$, $y_{\min} = \min(y_i)$.

Table 3.1 Measurement errors of the vision sensor in shaking table tests.

Frequency (Hz)	Amplitude(mm)	NRMSE
1.0	10	0.30%
5.0	2	1.40%
10.0	1	2.20%
20.0	1	3.40%

Based on Eq. (3.1), errors between the displacement time histories measured by the LVDT and the vision sensor are computed and tabulated in Table 3.1 for each of the sinusoidal excitations. As shown in the table, the error increases as the excitation frequency increases, and the vibration amplitude decreases. Despite the low amplitude of vibration (only 1 mm in the 10 and 20 Hz vibration tests), the vision sensor demonstrates high accuracy by achieving low measurement errors: 3.40% or less in terms of the NRMSE, validated by the high-accuracy LVDT reference sensor.

3.2 Shaking Table Test of Frame Structure 1

The performance of the computer vision sensor for simultaneous measurement of multiple points is evaluated through a shaking table test of a scaled three-story frame model structure in the Sensing, Monitoring, and Robotic Technology (SMaRT) Laboratory at Columbia University [31]. As shown in Figure 3.4, the aluminum frame structure is bolt-connected at all the column-floor connections.

The displacements measured by the computer vision sensor based on the OCM template-matching algorithm are compared with those based on the upsampled cross-correlation (UCC) algorithm. Both algorithms are described in Chapter 2. To simplify the expressions, UCC and OCM denote the vision sensor systems based on UCC and OCM, respectively.

3.2.1 Test Description

The frame structure is fixed on the shaking table (model APS113 by APS Dynamics Inc.), which is excited in the horizontal direction by white noise signals. Four predesigned black and white artificial target panels (each measuring 99 mm × 75 mm) are mounted on the structure. Meanwhile, four bolt connections are selected to study the performance of the vision sensor when tracking low-contrast natural targets on the structure. As references, the displacements are also

(a)   (b)

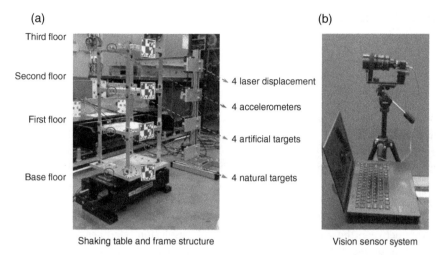

Third floor

Second floor   ⊁ 4 laser displacement

                         ⊁ 4 accelerometers

First floor

                         ⊁ 4 artificial targets

Base floor   ⊁ 4 natural targets

Shaking table and frame structure   Vision sensor system

Figure 3.4 Shaking table test of a three-story frame structure: (a) shakinq table and frame structure; (b) vision sensor system. *Source:* Reproduced with permission of John Wiley & Sons.

measured by four high-resolution, high-accuracy laser displacement sensors (LDSs; (model LK-G407 by KEYENCE), which are installed at stationary points to measure displacements of the floors and the base of the frame structure.

The camera of the vision senor system is placed 8 m from the structure. During the measurements, the 640 × 512 pixel, 8-bit grayscale video images captured by the camera are streamed into the computer through a USB 3.0 cable. In this test, four small areas from the artificial targets and the four bolt-connection areas on all the floors are simultaneously registered as templates in a signal video frame taken by the single camera. A total of 5.6 ms is needed to process each video frame, including the time to read, prepare, template-match, and display the image. Thus, it is possible to achieve real-time simultaneous measurement of displacements at eight target points with a sampling rate of 150 fps.

3.2.2 Subpixel Resolution

Pixel-level template matching may result in unacceptable measurement errors if the displacement to be measured has the same order of magnitude as the scaling factor. This happens when using a single camera to measure displacements of multiple points spaced over a large structure. In this case, the subpixel technique can be adopted to make template matching fall at a fractional pixel location [28].

To better understand how the subpixel technique improves measurement accuracy, displacements extracted from video images by tracking the artificial target

Table 3.2 Different levels of subpixel resolution.

Subpixel (pixel)	1	0.5	0.2	0.05
Resolution (mm)	±0.669	±0.335	±0.134	±0.034

on the base floor are used as a demonstration. Four subpixel levels – 1, 0.5, 0.2, and 0.05 pixels – are chosen, with the corresponding resolutions tabulated in Table 3.2. Recall again that for the vision sensor based on UCC, a desired subpixel resolution can be easily achieved by simply adjusting the upsampling factor. In this testing, the scaling factor is 1.338 mm/pixel, providing ±0.669 mm resolution.

In Figure 3.5a, at the integer-pixel resolution (1.338 mm), the displacement errors between the vision sensor and the high-resolution reference LDSs can be clearly observed. On the other hand, after employing different levels of the sub-pixel technique, the displacement measured by the vision sensor agrees better with that by the LDSs as the resolution improves, with NRMSE errors in Eq. (3.1) dropping from 6.41% to 3.80%, 1.73%, and 1.35% respectively for the zoomed-in segments in Figure 3.5.

Note that in an ideal environment where video images suffer no distortion or noise, a larger upsampling factor would yield smaller measurement errors. However, the subpixel accuracies reported in many other studies vary within orders of magnitude from 0.5 to 0.01 pixels [35, 36], as images may be contaminated by various external environmental noises and system noises arising from the electronics of the imaging digitizer. For the following tests in Section 3.2, $\kappa = 20$ is selected for the UCC-based vision sensor.

3.2.3 Performance When Tracking Artificial Targets

To evaluate the performance of the vision sensors based on the UCC and OCM template-matching algorithms, displacements are measured by tracking both high-contrast artificial targets and low-contrast natural targets (i.e. bolt connections) and comparing them with those from the high-accuracy reference LDSs. The measurements are respectively referred to as UCC (artificial target), OCM (artificial target), UCC (natural target), OCM (natural target), and LDS. The performance of the UCC and OCD algorithms is also compared.

Figure 3.6a plots the displacement time histories of the base floor by the vision sensors tracking the artificial targets based on OCM and UCC, respectively, and by the LDS reference sensor. Excellent agreement is observed among the three measurements. In Figure 3.6b–d, the relative displacements of the first, second, and third floors with respect to the base floor displacement are plotted in enlarged

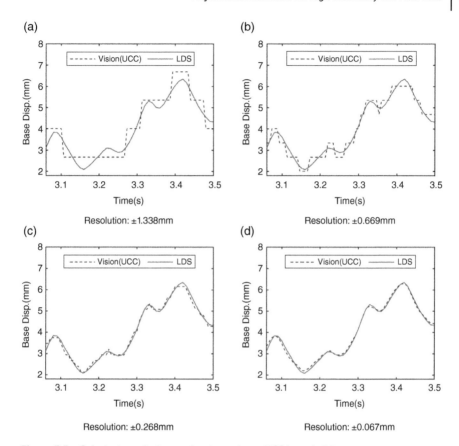

Figure 3.5 Subpixel resolution evaluation using a UCC-based vision sensor: (a) resolution: ±1.338 mm; (b) resolution: ±0.669 mm; (c) resolution: ±0.268 mm; (d) resolution: ±0.067 mm.

time segments between 1 and 5.5 seconds for better illustration. Note that the scaling factor for the vision sensor in this test is 1.338 mm/pixel, meaning the expected resolution is 0.669 mm from pixel-level template matching. However, by implementing the subpixel technique for both OCM and UCC, the measurement resolution of the vision sensors can be significantly improved so that low-amplitude displacements ranging from 0 to 3 mm can be accurately measured.

3.2.4 Performance When Tracking Natural Targets

Instead of using artificial targets, existing natural targets – i.e. the four bolt connections on the frame structure shown in Figure 3.4 – are used as tracking targets. Compared with the results using artificial targets, the discrepancies between the

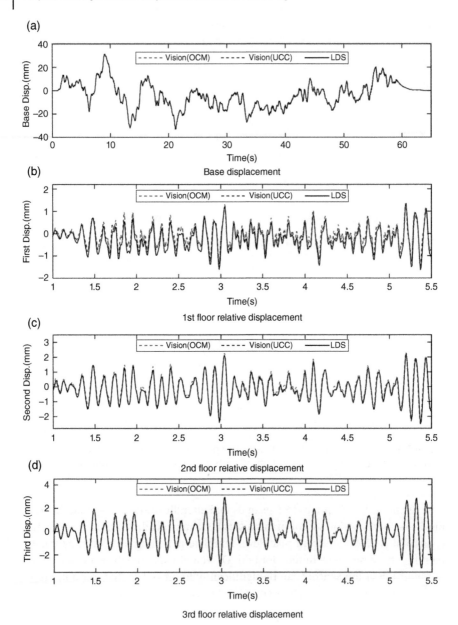

Figure 3.6 Comparison of displacements by OCM (artificial target), UCC (artificial target) and LDS: (a) base displacement; (b) first-floor relative displacement; (c) second-floor relative displacement; (d) third-floor relative displacement. *Source:* Reproduced with permission of John Wiley & Sons.

relative displacement of each floor measured by the vision sensor and the LDSs are slightly increased when using natural targets for the vision sensor. This can be observed by comparing Figure 3.7b–d with Figure 3.6b–d. It is mainly attributable to the greater difficulty in tracking the natural targets because their contrast is lower than that of the artificial targets. It is expected that the errors will decrease as the amplitude of the relative displacement increases.

By tracking natural targets without requiring artificial targets to be installed on fixed locations of the structure, the vision sensor provides a unique advantage for measuring structures that are difficult to access, compared with conventional sensors that must be installed on the structure. Furthermore, it is possible to alter the target points from a single video measurement using post-processing, which cannot be done with conventional on-structure sensors.

3.2.5 Error Quantification

Table 3.3 tabulates the measurement NRMSE errors for each floor displacement of the frame structure computed from Eq. (3.1). Note that these are the absolute displacements for the base, first, second, and third floors. Both the OCM and UCC vision sensors demonstrate a high measurement accuracy with a maximum NRMSE error of 0.72%.

3.2.6 Evaluation of OCM and UCC Robustness

In realistic field environments, unfavorable conditions such as fluctuating illumination, partial target occlusion, and background disturbances are often encountered, which degrade the image quality and thus the robustness of template matching and measurement accuracy. To investigate the robustness of the UCC- and OCM-based vision sensors under harsh environmental conditions, the following four cases are designed and tests conducted to measure displacements of the frame structure under sinusoidal shaking table excitations [31]. The test setup is shown in Figure 3.8:

- Case 1: Measurement by tracking an artificial target in normal light
- Case 2: Measurement by tracking a natural target (bolt connection) in normal light
- Case 3: Measurement by tracking the bolt connection in dim light that simulates fluctuating illumination and background disturbance
- Case 4: Measurement by tracking a partially obscured bolt connection to simulate partial template occlusion

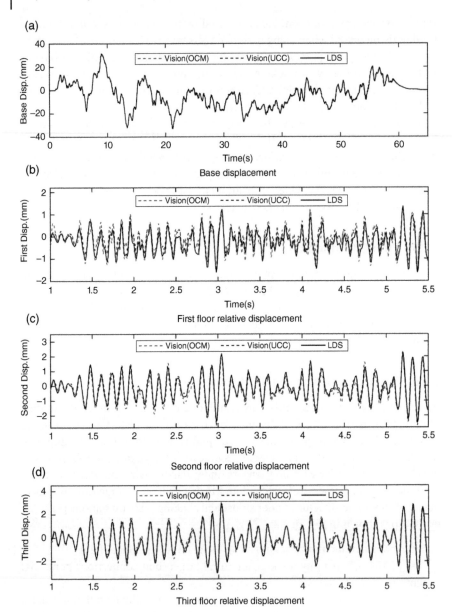

Figure 3.7 Comparison of displacements by OCM (natural target), UCC (natural target), and LDS: (a) base displacement; (b) first-floor relative displacement; (c) second-floor relative displacement; (d) third-floor relative displacement. *Source:* Reproduced with permission of John Wiley & Sons.

Table 3.3 Measurement errors: NRMSE (%).

Floor	Vision (OCM)		Vision (UCC)	
	Artificial target	Natural target	Artificial target	Natural target
Base	0.35	0.72	0.39	0.60
First	0.24	0.56	0.28	0.45
Second	0.21	0.43	0.27	0.35
Third	0.14	0.35	0.18	0.32

Source: Reproduced with permission of John Wiley & Sons.

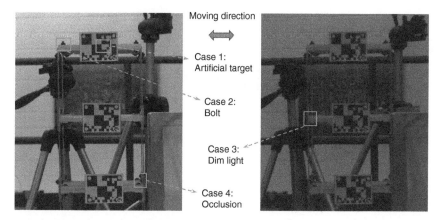

Figure 3.8 Evaluation of robustness in unfavorable conditions. *Source:* Reproduced with permission of John Wiley & Sons.

In normal light, both OCM and UCC can accurately measure dynamic displacements. Figures 3.9 and 3.10 plot the displacements of the top floor of the frame under 1-Hz sinusoidal excitation, measured by tracking an artificial target (Case 1) and a natural target (bolt connection; Case 2) and processed with the OCM algorithm, in comparison with the UCC algorithm. Good agreements are observed, and both UCC and OCM accurately measure the displacements. In the contours of the UCC correlation function, maximum values can be located at the true template-matching positions, as shown in Figures 3.9b and 3.10b for an example time of 2.5 seconds.

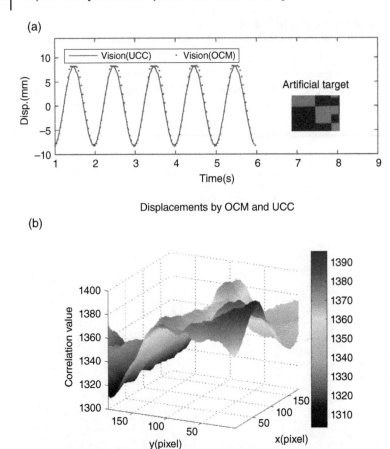

Figure 3.9 Case 1 comparison: (a) displacements by OCM and UCC; (b) UCC cross-correlation function contour. *Source:* Reproduced with permission of John Wiley & Sons.

In dim light, as in Case 3, OCM can successfully measure displacement but UCC completely fails to do so, as shown in Figure 3.11a, The contour of the UCC correlation function, shown at an example time of 2.5 seconds in Figure 3.11b, presents a sequence of peaks, making it impossible to find the right template-matching location.

When the target is partially occluded, as in Case 4, in which part of the template at the bottom-right corner of the frame is occluded when the structure is moving to the right, OCM successfully measures the displacements, while UCC completely fails to track the template. The measured time histories of these two template-matching algorithms are shown in Figure 3.12a. Figure 3.12b is

(a)

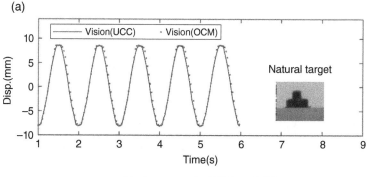

Displacements by OCM and UCC

(b)

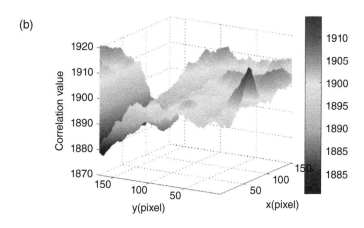

UCC cross correlation function contour

Figure 3.10 Case 2 comparison: (a) displacements by OCM and UCC; (b) UCC cross-correlation function contour. *Source:* Reproduced with permission of John Wiley & Sons.

the contour of the UCC correlation function at an example time of 2.0 seconds: multiple peaks appear, whose maximum is not located at the right matching point.

In summary, the OCM vision sensor is robust in unfavorable conditions such as dim light and partial image occlusion, which are challenges commonly encountered in outdoor field measurements. This is reasonable since OCM employs gradient information in the form of orientation codes that are inherently invariant to variations in image intensity and thus robust when irregularities are present. On the other hand, the UCC vision sensor utilizes image intensity values for template matching and thus has difficulty dealing with unfavorable conditions.

(a)

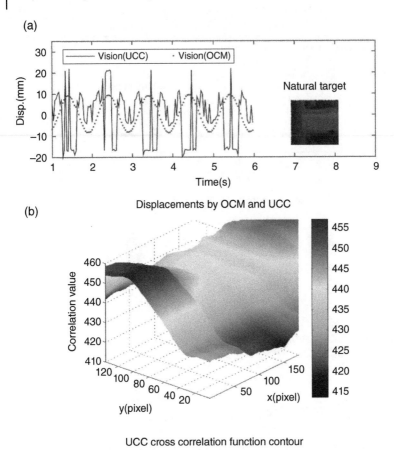

Time(s)

(b)

Displacements by OCM and UCC

Correlation value

UCC cross correlation function contour

Figure 3.11 Case 3 comparison: (a) displacements by OCM and UCC; (b) UCC cross-correlation function contour. *Source:* Reproduced with permission of John Wiley & Sons.

Since the gradient-based OCM template-matching algorithm is more robust than the conventional intensity-based UCC algorithm, the computer vision sensor based on the OCM algorithm is adopted throughout the remainder of the book.

3.3 Seismic Shaking Table Test of Frame Structure 2

The performance of the vision sensor is further evaluated through seismic shaking table tests in the Civil Engineering Laboratory, University of California, Irvine. The vision sensor is used to measure the seismic response of a three-story steel frame building model excited by a biaxial hydraulic seismic shaking table. The

(a)

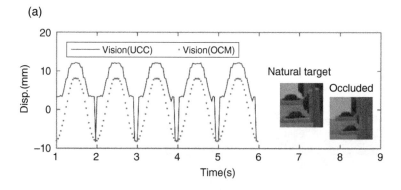

Displacements by OCM and UCC;

(b)

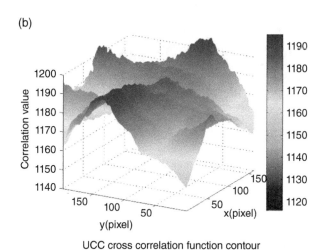

UCC cross correlation function contour

Figure 3.12 Case 4 comparison: (a) displacements by OCM and UCC; (b) UCC cross-correlation function contour. *Source:* Reproduced with permission of John Wiley & Sons.

table is $10' \times 12'$, and the three-story steel building frame model is fixed on the table, as shown in Figure 3.13.

The overall floorplan dimensions of the model are $60'' \times 60''$, and each story is $30''$ high. Two seismic ground motion records are selected – the 1995 Great Hanshin-Awaji earthquake and the 2000 Western Tottori earthquake – as they represent distinctively different frequency content. The response displacement of the model is measured by the vision sensor at 30 fps and an LVDT (model P-20A, Sensors & Controls Corporation) with a sampling rate of 125 Hz as a reference sensor. The target panel has 125 mm between two points horizontally and 100 mm

Figure 3.13 A steel building frame model on a seismic shaking table.
Source: Reproduced with permission of John Wiley & Sons.

between two points vertically. The LVDT is installed on the top-front side of the model, and a digital camcorder (Panasonic PV-GS35) with a telescopic lens (8×) is placed 5 m from the target panel, as shown in Figure 3.14.

Figure 3.15 shows the seismic response displacements of the building frame, together with their power spectral densities, measured by the two sensors under the two ground motions. As shown in Figure 3.15a, the displacement time histories measured by the LVDT and the vision sensor are in excellent agreement with each other for both ground motions. The root mean square error (RMSE) values between the displacement time histories measured by the LVDT and the vision sensor were 0.671 mm for the Great Hanshin-Awaji earthquake and 0.660 mm for the Western Tottori Earthquake, respectively. As shown in the power spectral densities in Figure 3.15b, the frequency domain responses are also in excellent agreement. The RMSE values between these two frequency domain responses are $0.355 \, mm^2/Hz$ for the Great Hanshin-Awaji earthquake simulation and $0.188 \, mm^2/Hz$ for excitation by the Western Tottori earthquake simulation, respectively. The shaking table test results demonstrate the capability of the vision sensor to measure structural responses to actual earthquakes with a wide range of frequencies.

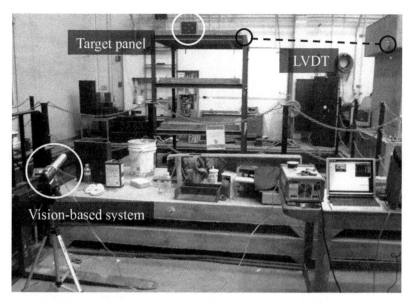

Figure 3.14 Seismic shaking table setup. *Source:* Reproduced with permission of John Wiley & Sons.

3.4 Free Vibration Test of a Beam Structure

To evaluate the accuracy of the practical calibration method, free vibration tests are conducted on a simply supported beam model. This beam is also used for modal testing, which will be presented in Chapter 4. The CMOS camera listed in Table 2.1 is used. The optical lens has a focal length varying from 16 to 160 mm with manual focus.

3.4.1 Test Description

As shown in Figure 3.16, the simply supported beam model is made of aluminum. A pin support at the left end allows the beam to rotate but not to translate either vertically or horizontally. A roller support at the right end allows the beam to rotate and translate longitudinally.

As illustrated in Figure 3.17, 30 small black dots, numbered 2 through 31, are marked along the side of the beam as targets for motion tracking. During the measurement, 1280 × 240 pixel, 8-bit grayscale video images captured by the camera are streamed into the computer through a USB 3.0 cable with a sampling rate of 50 fps.

As references, the displacements are also measured by two high-accuracy LDSs (model LK-G407 by KEYENCE) at point 9 and point 16, respectively. In addition, six accelerometers (model W352C67 by PCB Piezotronics Inc.) are installed on the

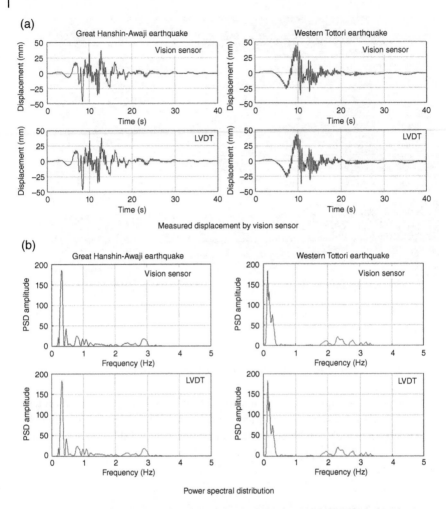

Figure 3.15 Experimental results of the seismic shaking table test: (a) measured displacement by the vision sensor; (b) power spectral distribution. *Source:* Reproduced with permission of John Wiley & Sons.

beam to further compare the experimental modal analysis results (discussed in Chapter 4). Both the LDSs and accelerometers use a sampling frequency of 50 Hz.

3.4.2 Evaluation of the Practical Calibration Method

As introduced in Chapter 2, in order to obtain structural displacements from the captured video images, the establishment of the relationship between the pixel coordinates and the physical coordinates (e.g. the scaling factor in units of mm/pixel)

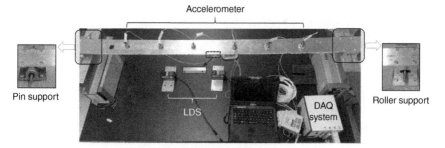

Figure 3.16   Test setup for the simply supported beam.

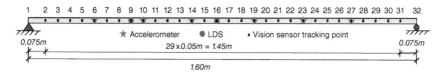

Figure 3.17   Schematic of sensor placement. *Source:* Reproduced with permission of Elsevier.

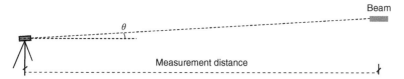

Figure 3.18   Case of a non-perpendicular camera optical lens axis. *Source:* Reproduced with permission of Elsevier.

is required. When the camera's optical axis is perpendicular to the object surface, all points on this surface have equal depth of field, which means these points can be equally scaled down into the image plane. In this case, only one identical scaling factor is needed. This book presents a practical calibration method in which the scaling factor is obtained based on the known physical dimensions of the object surface and its corresponding image dimensions in pixels, as expressed in Eq. (2.3).

In realistic tests, it is often impossible to avoid tilting the camera's optical axis by a small angle θ in order to track the measured object surface, as illustrated in Figure 3.18. The theoretical studies in Chapter 2 reveal that the error increases as the camera tilt angle increases. Here, the beam model shown in Figure 3.16 is used to investigate the accuracy of scaling factor SF_1 given a camera tilt angle.

To better illustrate this, the known vertical length (109 mm) of a predesigned marker panel is used to estimate the vertical scaling factors for testing three camera tilt angles: 3°, 5°, and 9°. As shown in Figure 3.19, the tilt angle of the camera reduces the vertical length of the marker panel in the image, resulting in

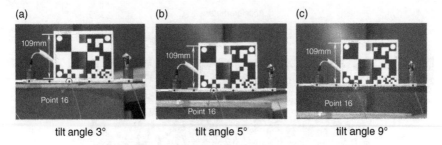

Figure 3.19 Images of a marker panel for different camera tilt angles: (a) 3°; (b) 5°; (c) 9°. *Source:* Reproduced with permission of Elsevier.

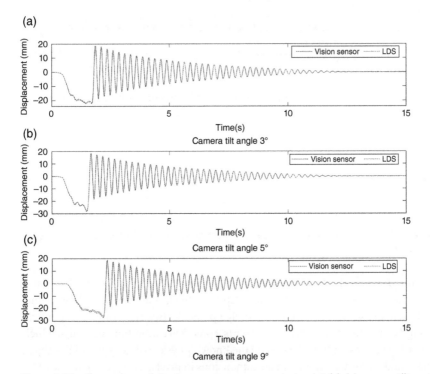

Figure 3.20 Comparison of displacement measurement at point 16: (a) camera tilt angle 3°; (b) camera tilt angle 5°; (c) camera tilt angle 9°.

a different scaling factor value. At the three tilt angles 3°, 5°, and 9°, the scaling factors are 0.394, 0.407, and 0.426 mm/pixel, respectively.

Free vibration of the beam is generated by inducing an initial displacement at point 4. The beam displacement time history at point 16 is measured by the computer vision sensor by tracking the corresponding black dot and compared with that from the LDS, as shown in Figure 3.20. The NRMSEs of the vision sensor corresponding to the three camera tilt angles are 1.0, 1.2, and 1.4%, respectively. This demonstrates that the estimated scaling factors from utilizing known physical dimensions can achieve satisfactory measurement accuracy even when the camera's optical axis is not perfectly perpendicular to the plane of the object motion.

Note that uncertainties exist when manually picking the image dimensions using a mouse. It is recommended to use the mean value from several repeated picking operations to average out some of the random error. Moreover, when a series of targets are tracked to measure displacements at multiple points along the structure, due to projective distortion, different scaling factors should be determined by using known physical dimensions closer to or encompassing each of the targets. If such dimensions are not available, predesigned marker panels, as demonstrated in this test, can be attached at those locations as references. To minimize the estimation error of the scaling factors, measures should be taken to guarantee that the marker panels are placed in the same plane as the object surface.

3.5 Field Test of a Pedestrian Bridge

In order to evaluate the efficacy of the computer vision sensor in measuring the dynamic response of an actual structure, field tests are conducted on the Streicker Bridge, a pedestrian bridge located on the Princeton University campus [28]. The bridge has a main span and four approaching legs. The main span is a deck-stiffened arch, and the legs are curved continuous girders supported by steel columns. Two sets of dynamic loading tests are carried out on the third span of the southeast leg. As shown in Figure 3.21, one artificial target panel is installed on the mid-span of the bridge. The camera is located about 40 m from the target panel, and the camera's optical axis is tilted by approximately 15° with respect to the normal direction of the bridge surface. The scaling factor is obtained using the practical method in Eq. (2.3). Due to the long distance between the ground and the bridge's bottom surface, it is very difficult to install a reference LVDT or LDS. Instead, one accelerometer (model W352C67 by PCB Piezotronics Inc.) is installed on the mid-span of the bridge, next to the target panel, as a reference sensor.

Two types of loading excitations are applied to the bridge by a group of pedestrians. The first type attempts to simulate dynamic loads with broadband frequency contents. The pedestrians ran on the bridge deck randomly with varying speeds, rhythms, and directions and no particular pattern. Figure 3.22 plots the vertical displacement time history of the bridge measured by the vision sensor, together

(a)

Figure 3.21 Field test: (a) Streicker Bridge; (b) artificial target.

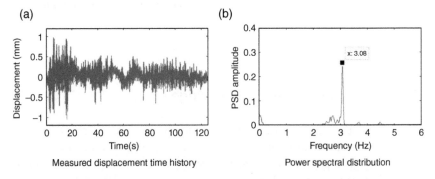

Measured displacement time history

Power spectral distribution

Figure 3.22 Randomly running pedestrians: displacement measurement by vision sensor: (a) measured displacement time history; (b) power spectral distribution.

with the corresponding PSD amplitude. Figure 3.23 plots the acceleration time history from the accelerometer and the corresponding PSD amplitude. Both power spectral density (PSD) plots show the dominant frequency of the bridge to be 3.08 Hz, and the two higher frequencies are 3.68 and 4.47 Hz. This confirms that the computer vision sensor system can achieve the same dynamic measurement performance as the conventional accelerometer in the field.

The second type of loading attempts to excite the first vibration mode of the bridge. The pedestrian participants jumped in unison on the mid-span of the bridge deck with a frequency around 3 Hz, which is close to the estimated first natural frequency of the bridge. Figures 3.24 and 3.25 plot the displacement and acceleration time histories obtained from the vision sensor and the accelerometer, respectively, together with their corresponding PSD results. As expected, only the

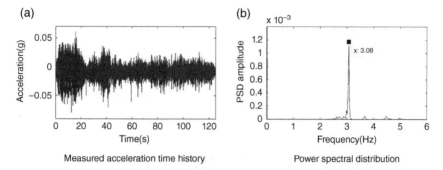

Figure 3.23 Randomly running pedestrians: acceleration measurement: (a) measured acceleration time history; (b) power spectral distribution.

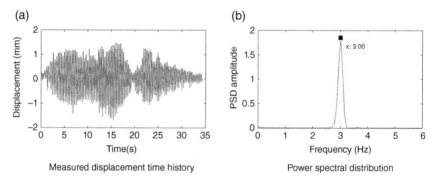

Figure 3.24 Jumping pedestrians: displacement measurement by vision sensor: (a) measured displacement time history; (b) power spectral distribution.

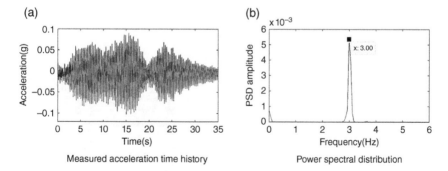

Figure 3.25 Jumping pedestrians: acceleration measurement: (a) measured acceleration time history; (b) power spectral distribution.

first mode of vibration is induced. Again, the identified first frequencies based on the two sensors show a perfect match.

From Figures 3.22–3.24, it is observed that the amplitude of the mid-span bridge response displacement is less than 2 mm. The computer vision sensor successfully measures the low-amplitude displacement. These displacement time histories are produced in real time. In summary, the efficacy of the computer vision sensor for structural dynamic response measurements is validated in this field test.

3.6 Field Test of a Highway Bridge

The computer vision sensor is further tested on a typical single-span stiff highway bridge, which deforms little under moving vehicle loads. The Samseung Bridge is a single-span highway bridge consisting of five steel plate girders, as shown in Figure 3.26. This is one of the three bridges built by the Korea highway corporation on a 7.7 km test road (next to an ordinary two-lane expressway along the

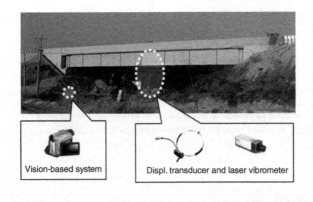

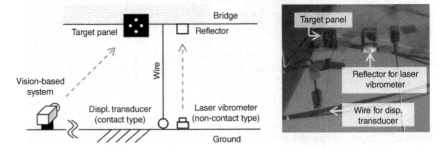

Figure 3.26 Field test on a highway bridge. *Source:* Reproduced with permission of John Wiley & Sons.

Joongbu Inland Expressway in Korea) to verify and enhance the pavement design guides based on measured data under real traffic and environmental conditions. The bridge's span length is 40 m. Moving vehicle tests are performed using three dump trucks weighing 15, 30, and 40 tons and running at speeds of 3 and 50 km/h.

The experimental setup is shown in Figure 3.26. The vertical response displacement of the bridge at the mid-span is measured by the computer vision sensor at a sampling rate of 30 fps. In addition, two types of reference sensors are used: a contact type conventional displacement transducer (i.e. a string potentiometer) at a sampling rate of 1000 Hz (OU displacement transducer, Tokyo Sokki Kenkyujo Co. Ltd.) and a noncontact type laser vibrometer at a sampling rate of 100 Hz (OFV-505 Standard Optic Sensor Head and OFV-5000 Modular Controller, Polytec, Inc.). The contact-type LVDT is installed on the ground (as the stationary reference point) and directly wired to the mid-span measurement point on the bridge. The noncontact laser vibrometer is also installed on the ground below the bridge, pointing at the same measurement point, and a reflector is installed at the measurement point to reflect the emitted laser. In contrast, a camera with a telescopic lens is placed 20 m from the target and captured an image 3 cm (480 pixels) vertically (0.00625 mm/pixel). A target panel is installed at the measurement point.

Figure 3.27 compares the displacement time histories measured by these three different types of sensors under the moving trucks with different weights and speeds. At a speed of 3 km/h, the maximum vertical displacements of the bridge mid-span are approximately 1.0, 2.0, and 2.5 mm for the 15, 30, and 40-ton vehicles, respectively. Among the three measurement systems, the conventional displacement transducer shows the highest measurement noise. For the low-speed vehicles, as shown in Figure 3.27a, the displacement (mainly static displacement) measured by the vision-based system is very close to those from the highly accurate laser vibrometer. For the high-speed (50 km/h) vehicles, as shown in Figure 3.27b, the displacements measured by the vision sensor system are approximately 10% lower than those measured by the laser vibrometer. This may be due to the fact that with the high-speed truck induced vibration of the bridge, the image acquisition speed of 30 fps was insufficient to capture the bridge's high-frequency dynamic vibration.

3.7 Field Test of Two Railway Bridges

As a railway bridge ages, deteriorates, and loses its stiffness, its deflections under live trainloads increases, which can potentially cause track instability and loss of contact between the rail and train wheels. Excessive deflections also accelerate fatigue in bridge structures, causing significant safety concerns. Therefore, deflection is considered one of the most important quantities to be closely monitored for

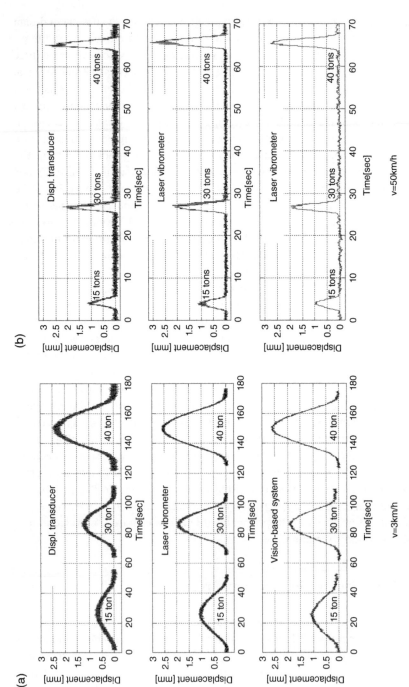

Figure 3.27 Experimental results of field tests on a highway bridge: (a) v = 3 km/h; (b) v = 50 km/h. *Source:* Reproduced with permission of John Wiley & Sons.

Figure 3.28 View of the two testbed bridges. *Source:* Reproduced with permission of ASCE Library.

a railroad bridge. A survey of North American railroad bridge structural engineers revealed a consensus that the top research priority is to measure real-time bridge displacements under trainloads, which could provide quantitative information for bridge maintenance [37].

In collaboration with the Transportation Technology Center, Inc. (TTCI) in Pueblo, Colorado, field tests of the computer vision sensor system are performed during the day and at night to remotely measure the dynamic displacement of two TTCI testbed bridges. Figure 3.28 shows the two bridges: one is a state-of-the-art hybrid composite beam (HCB) girder bridge, and the other is a 100-year old riveted steel girder bridge. Each of the two spans of the steel girder bridge is 16.9 m long, and the main span of the HCB bridge is 12.8 m long. The field tests focus on measuring the trainload-induced mid-span vertical displacements of these two bridges and the comparative performance of the vision-based sensor with and without artificial target panels and a conventional LVDT for moving trainloads during the day and at night.

3.7.1 Test Description

For the field tests, two identical CMOS cameras are used to track an artificial target on the bridge and a nearby natural target, respectively. Each camera is equipped with a 160 mm lens and connected to a notebook computer with the real-time OCM software package installed. As shown in Figure 3.29, each camera is fixed on a tripod 9.15–60.96 m from the bridge. The artificial target panel is installed on the mid-span measurement point of the bridge. Light-emitting diode (LED) lights are also installed on the bridge as targets for the measurements at night. Figure 3.30 shows the target panel and various natural surface features, including rivets and letters, and the LED lights on the bridge, which are used as measurement targets. As a reference sensor, a conventional contact-type displacement sensor, an LVDT, is installed on the mid-span of the bridge with one end wired to a stationary reference point on the ground using a string.

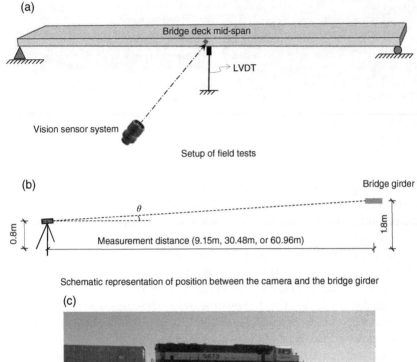

(a)

Bridge deck mid-span

LVDT

Vision sensor system

Setup of field tests

(b)

Bridge girder

θ

0.8m

1.8m

Measurement distance (9.15m, 30.48m, or 60.96m)

Schematic representation of position between the camera and the bridge girder

(c)

Remote measurement of bridge displacement under moving trainloads

Figure 3.29 Field tests on the railway bridge: (a) setup of field tests; (b) schematic representation of the position between the camera and the bridge girders (c) remote measurement of bridge displacement under moving trainloads. *Source:* Reproduced with permission of ASCE Library.

During the two-day tests, 41 sets of bridge displacement time histories under trainloads are measured in various conditions. Representative measurements are presented here, including four measurements made during the day on the HCB bridge and four at night on the steel bridge with a flashlight or using the LED target. The test parameters for the selected measurements are summarized in

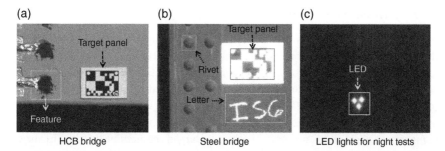

(a) HCB bridge (b) Steel bridge (c) LED lights for night tests

Figure 3.30 Target panel and existing features on the railway bridges: (a) HCB bridge; (b) steel bridge; (c) LED lights for night tests. *Source:* Reproduced with permission of ASCE Library.

Table 3.4 Test conditions of eight representative measurements.

Bridge	Test	Measurement distance (m)	Train speed (km/h)	Measurement target	Scaling factor (mm/pixel)
HCB bridge (day)	H1	9.15	40.23	Panel and feature	0.94
	H2	30.48	40.23	Panel and feature	1.90
	H3	30.48	64.36	Panel and feature	1.88
	H4	60.96	64.36	Panel and feature	3.83
Steel bridge (night)	S1	9.15	64.36	Panel	0.83
	S2	9.15	64.36	Feature (rivet)	0.78
	S3	9.15	64.36	Feature (letter)	1.20
	S4	9.15	64.36	LED lights	1.20

Table 3.4. The scaling factors are obtained based on the known dimensions of the artificial target panels. For field measurements without using the target panel, the known dimensions of existing features on the structure, such as the size of the nuts and rivets known from design drawings, can be used to obtain the scaling factor.

Two freight trains are used for the field tests. The train used for the daytime measurement – tests H1–H4 – has one locomotive weighing 1.75×10^6 N and 15 cars weighing 1.40×10^6 N each. The train used for the night measurement – tests S1–S4 – has three locomotives weighing 1.75×10^6 N each and 105 cars weighing 1.40×10^6 N each.

3.7.2 Daytime Measurements

Measurements are taken during the day on the HCB bridge. Figures 3.31–3.34 plot the displacement time histories measured in tests H1–H4 by the three sensor systems: the LVDT and the two vision sensors – one targeting the preinstalled artificial target panel (called Vision (Target)) and the other the natural surface feature (called Vision (Feature)). The measurement distance varies, at 9.15, 30.48, and 60.96 m; and the train speeds are 40.23 and 64.36 km/h. For a better view of the comparison, each plot is presented with an enlarged segment of the time history. In general, the measurements from the three sensors agree well.

3.7.3 Nighttime Measurements

To investigate the potential for using the vision-based sensor system at night, measurements S1–S4 are taken on the steel bridge. Test S1 is conducted at night by focusing the camera on the preinstalled target panel shown in Figure 3.30b, which is illuminated by a flashlight. Figure 3.35 plots the displacement time histories measured in test S1 by the LVDT and the vision sensor. Again, one segment of the time history for each measurement is enlarged for a better view. Tests S2 and S3 are

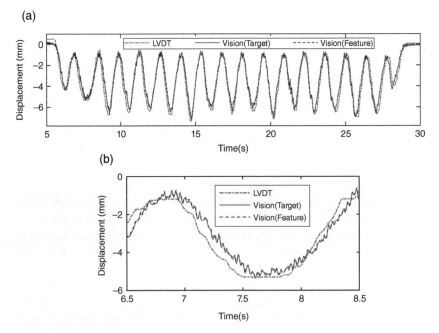

Figure 3.31 Test H1: comparison of displacements by three sensors (day). *Source:* Reproduced with permission of ASCE Library.

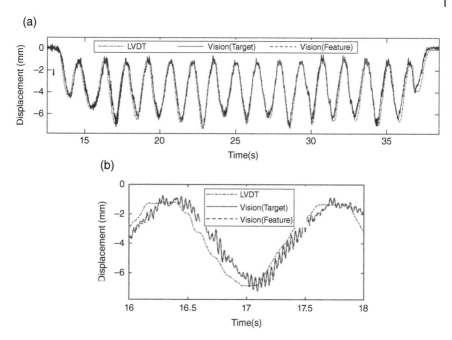

Figure 3.32 Test H2: comparison of displacements by three sensors (day).
Source: Reproduced with permission of ASCE Library.

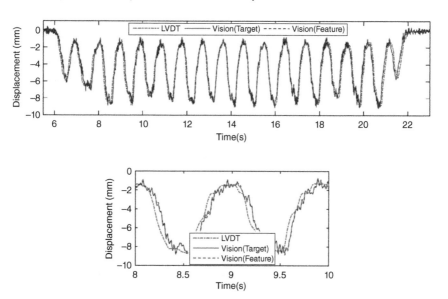

Figure 3.33 Test H3: comparison of displacements by three sensors (day).
Source: Reproduced with permission of ASCE Library.

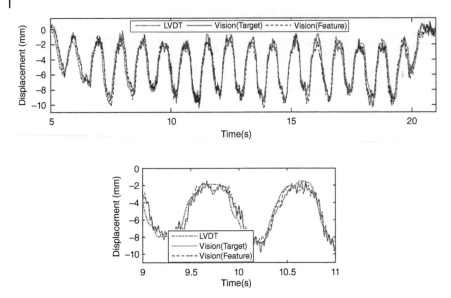

Figure 3.34 Test H4: comparison of displacements by three sensors (day).
Source: Reproduced with permission of ASCE Library.

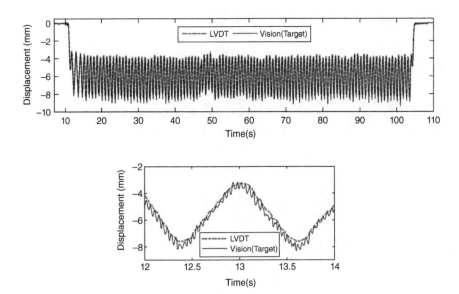

Figure 3.35 Test S1: comparison of displacements by two sensors (night).
Source: Reproduced with permission of ASCE Library.

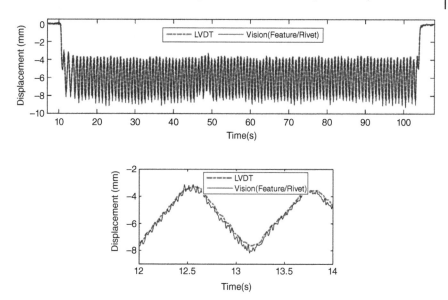

Figure 3.36 Test S2: comparison of displacements by two sensors (night).
Source: Reproduced with permission of ASCE Library.

carried out at night by focusing the camera on the rivets or letters shown in Figure 3.30b, which are also illuminated by a flashlight. The displacement time histories measured by the vision sensor and the LVDT are shown in Figures 3.36 and 3.37, respectively. Finally, test S4 is carried out at night using the preinstalled LED lights, shown in Figure 3.30c, without any additional illumination. Figure 3.38 compares the measured displacement time history with that from the LVDT. The train used for the night tests includes 3 locomotives and 105 cars.

As shown in these plots, the displacement time histories measured at night for the various test conditions agree very well. This proves the feasibility of using the vision-based sensor system for night measurements as long as a low-power light source such as a flashlight is available to illuminate an existing bridge surface feature as a tracking target or an LED light can be preinstalled on the bridge.

3.7.4   Field Performance Evaluation

The performance of the vision-based displacement sensor system is evaluated based on the results of the field test of the two railway bridges. The measurement error of the vision sensor is quantified by comparing the displacements measured by the two vision sensors (using artificial and natural targets, respectively) and the LVDT. Additional observations are made regarding the effects of various test parameters.

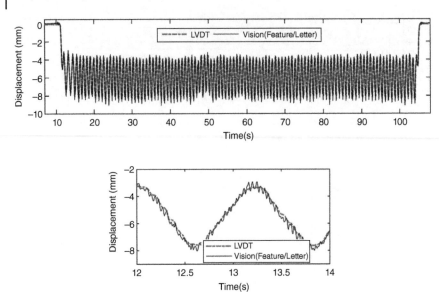

Figure 3.37 Test S3: comparison of displacements (night). *Source:* Reproduced with permission of ASCE Library.

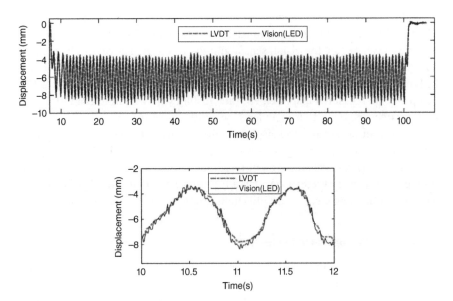

Figure 3.38 Test S4: comparison of displacements (night). *Source:* Reproduced with permission of ASCE Library.

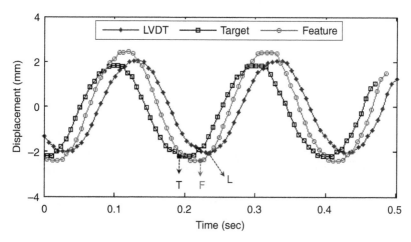

Figure 3.39 Schematic illustration of the displacement peak. *Source:* Reproduced with permission of ASCE Library.

As can be seen from the displacement time history plots from tests H1–H4 in Figures 3.31–3.34, there is a random time lag between displacements measured by the two vision sensors and the LVDT, despite the synchronized starting time for each measurement. The time lag is considered to be caused by two factors. First, the wind-induced sway of the LVDT string (between the bridge and the ground) may have caused a measurement error. It is observed that the wind is stronger during the H1–H4 tests in the daytime than during the S1–S4 tests at night. Second, the vision sensor uses the PC's internal clock for data acquisition, while the LVDT has its own data-acquisition device with its own internal clock. These clocks have their own errors. Therefore, it would be unreasonable to use point-to-point error-evaluation methods (e.g. NRMSE) to quantify measurement errors.

Instead, the measurement accuracy of the vision sensor is evaluated by comparing the peak values in each cycle of the oscillation in the displacement time histories. Figure 3.39 shows those displacement peaks: L (measured by the LVDT), T (by the vision sensor using an artificial target), and F (by the vision sensor using a bridge surface feature). Each train car's passage causes one cycle of bridge oscillation. For the HCB bridge, the single locomotive and 15 cars result in 17 displacement peaks, while for the steel bridge, the 3 locomotives and 105 cars yield 109 displacement peaks. The differences between the peaks measured by the three different sensors, defined as $(T-L)/L \times 100\%$, $(F-L)/L \times 100\%$, and $(F-T)/T \times 100\%$, are computed and plotted in Figures 3.40 and 3.41, respectively, for the tests on the HCB bridge and the steel bridge. Again, it is worth noting that the accuracy of the LVDT deteriorates due to wind-induced swaying of the string,

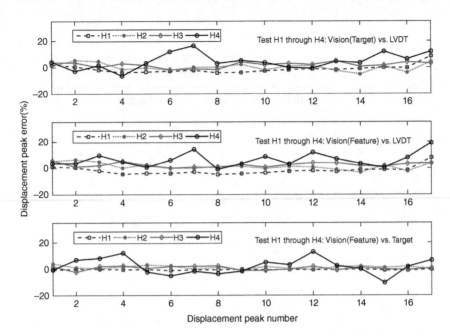

Figure 3.40 Errors between peak displacements of test H1–H4 of the HCB bridge.
Source: Reproduced with permission of ASCE Library.

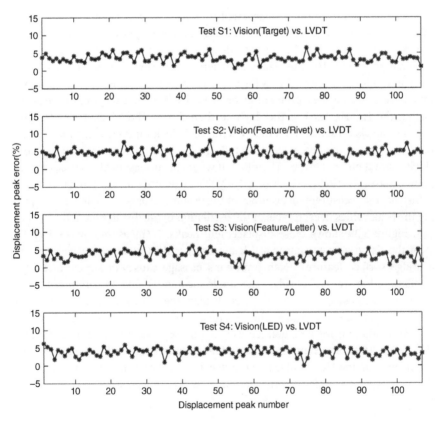

Figure 3.41 Errors between peak displacements of the steel bridge in tests S1–S4.
Source: Reproduced with permission of ASCE Library.

Table 3.5 Errors between peak displacements measured by different sensors.

Bridge type	Test	Measurement distance(m)	Maximum peak error (%)			Mean of absolute peak error (%)		
			T vs. L	F vs. L	F vs. T	T vs. L	F vs. L	F vs. T
HCB bridge	H1	9.15	4.10	4.82	1.05	2.20	2.95	0.60
	H2	30.48	5.20	6.50	2.83	2.27	2.19	1.39
	H3	30.48	4.82	5.70	2.40	2.43	2.43	1.39
	H4	60.96	16.42	19.16	12.91	5.71	6.40	4.95
Steel bridge	S1	9.15	6.45	N/A	N/A	3.57	N/A	N/A
	S2	9.15	N/A	7.98	N/A	N/A	4.48	N/A
	S3	9.15	N/A	7.18	N/A	N/A	3.37	N/A
	S4	9.15	N/A	6.53	N/A	N/A	3.79	N/A

especially during the HCB bridge tests, thus failing to provide true values of the bridge displacements for evaluating the vision sensor's accuracy.

Table 3.5 tabulates measurement errors computed from the displacement peaks, including the maximum differences between the displacement peaks measured by the three different sensors, and the corresponding mean values of the absolute differences between the displacement peaks in Figures 3.40 and 3.41.

Analyzing the railway bridge test results, the following observations can be made:

1) *Effect of the measurement distance.* Often, in the field, it is difficult to find a location close to the target structure to set up the camera, resulting in the need to measure the bridge displacement at a long distance. It is observed that the measurement error increases as the measurement distance increases. In field tests H1–H3, when the measurement distance varies from 9.15 to 30.48 m, the errors between the measurements by the different sensors, as shown in Table 3.5, are relatively small, with a maximum error in the range of 1.05– 6.50% and a mean error around 0.60–2.95%. In test H4, when the distance increases to 60.96 m, the measurement accuracy deteriorates, as shown in both the maximum error, in the range of 12.91–16.42%, and the mean error, which is 4.95–5.71%. Such errors related to the measurement distance can be caused by (i) camera vibration amplified by the zoom lens and (ii) heat haze.

2) *Accuracy of the vision sensors using artificial and natural targets.* From the results of tests H1–H3 shown in Figures 3.31–3.33 and Table 3.5 for column F vs. T, excellent agreement is observed between the measurements by the vision sensor with and without using the high-contrast artificial target panel. The

maximum error is 2.83%, and the corresponding mean error is only 1.39% for test H2 in which the measurement distance is 30.48 m. As the measurement distance doubles, the maximum error increases to 12.91% and the mean error to 4.95%, due to the increased difficulty of tracking the existing surface feature (which has a lower contrast than the artificial target panel). These errors may be acceptable, however, considering the benefit of avoiding the installation of a target panel on the bridge. In addition, without using an artificial target panel installed at a fixed location, the user has the flexibility to change the displacement measurement locations on the structure from a single video recording through post-processing. This flexibility provides a significant advantage for the vision sensor that cannot be matched by a conventional on-structure sensor.

3) *Night measurement.* A vision-based sensor requires light for measurement, which raises concerns about its usefulness for continuous 24/7 monitoring. The tests conducted at night show satisfactory measurement accuracy by the vision sensor when tracking a target illuminated with a dim light or LED lights. The results in Table 3.5 indicate that the maximum errors in all the measurements are approximately 6.45–7.98% in comparison with LVDT, with mean errors of 3.37–4.48%. The high measurement accuracy achieved in the night field tests demonstrates the robustness of the OCM algorithm in template matching from the images taken in dim light conditions.

4) *Limitations of the contact-type displacement sensor.* A number of difficulties associated with the LVDT are observed in this field test. First, it takes a long time to properly install the LVDT, due to the requirement to connect the sensor to a stationary reference point aligned in the direction of the displacement, as shown in Figure 3.28a. This could be highly challenging (if not impossible) for bridges crossing water or deep canyons. Second, the long string connecting the LVDT with the ground stationary point is subjected to wind vibration, which induces measurement errors.

5) *Limitations of the vision sensors.* Two problems associated with the vision sensors are recognized. First, camera vibrations caused primarily by ground motion induced by the moving train can affect measurement accuracy. In particular, when the zoom lens is used for the camera, it magnifies not only the images but also the camera vibration. A practical method will be presented in Section 3.10 to mitigate the camera vibration-induced measurement error. The second problem is heat haze, which occurs when the air is heated non-uniformly by a high ambient temperature during computer vision measurements. The non-uniformly heated air causes variation in its optical reflection index, resulting in image distortion. As the measurement distance increases, the resulting distortion between the target object and the lens of the camera becomes large, causing more measurement errors. This is a topic of future research.

3.8 Remote Measurement of the Vincent Thomas Bridge

Compared to accelerometers, displacement sensors are better suited for measuring the vibration of long-period structures such as long-span bridges. However, it is extremely difficult (if not impossible) to use conventional contact-type displacement sensors on long-span bridges, due to the lack of nearby stationary reference points. The computer vision sensor system is applied to two long-span suspension bridges to measure traffic-induced bridge vibration by tracking targets from hundreds of meters away. The first bridge is the Vincent Thomas Bridge, a 460 m suspension bridge that crosses Los Angeles Harbor in California. This field test is focused on evaluating the robustness of the computer vision sensor in long-distance measurements with or without using artificial targets under different lighting conditions.

As shown in Figure 3.42, two synchronized video cameras are placed at a stationary location approximately 300 m from the target at mid-span of the main bridge span. One camera focuses on an artificial target panel with random black and white patterns, which is installed at the mid-span, whereas the other camera targets existing rivets and edges on the bridge near the target panel.

Figure 3.43 shows the actual images captured by the two video cameras; one focuses on the installed target panel and the other on the existing rivet pattern close to the target panel. The resolution of the captured images is 640 × 480 pixels, and the frame rate is 60 fps. In this field test, vibration in the vertical direction of the bridge is measured. The scaling factor is obtained as 3.01 mm/pixel. Mid-span vertical displacements are measured in different lighting conditions. One measurement is in the morning in full sunlight, and the other is performed in the early

(a) (b)

Vincent Thomas Bridge (Los Angeles, CA) Test setup

Figure 3.42 Field test of the Vincent Thomas Bridge: (a) Vincent Thomas Bridge (Los Angeles, CA); (b) test setup. *Source:* Reproduced with permission of IEEE.

(a)   (b)

Artificial target panel   Natural rivet pattern

Figure 3.43 Actual images captured by two cameras: (a) artificial target panel; (b) natural rivet pattern. *Source:* Reproduced with permission of IEEE.

evening without sufficient illumination in order to evaluate the robustness of the OCM algorithm against varying light conditions.

Figure 3.44 compares the displacement time histories measured by focusing on the artificial target panel and the natural feature target in the morning and evening. The good agreement between the results of these two vision sensors demonstrates the effectiveness of the OCM-based vision sensor in tracking existing features from a distance of 300 m. Furthermore, the PSD amplitudes of the measured displacement time histories are plotted in Figure 3.45. Again, the displacements measured with and without the target panels agreed well in the frequency domain. A dominant frequency, 0.227 Hz, is identified from the measurement in the morning, and 0.229 Hz in the evening. This frequency is consistent with the bridge's fundamental frequency measured by accelerometers installed on the bridge in a previous study [32, 38].

Note that if only the structural frequency information is required from a dynamic test, the coordinate transformation is unnecessary. In other words, there is no need to determine the scaling factor to transform the pixel coordinate vibrations into physical coordinate vibrations, which would make the vision-based measurement procedure even more efficient.

3.9 Remote Measurement of the Manhattan Bridge

The second long-span bridge tested is the landmark Manhattan Bridge. Opened to traffic in 1909, this suspension bridge spans the East River in New York City, connecting Manhattan and Brooklyn. The main span is 448 m long. The bridge is 36.5 m wide, including 7 lanes in total and 4 subway lines, as shown in Figure 3.46.

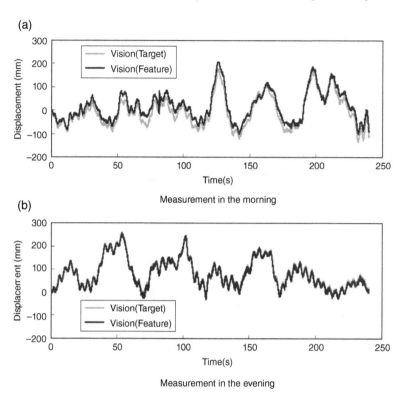

Figure 3.44 Displacement time histories: (a) measurement in the morning; (b) measurement in the evening. *Source:* Reproduced with permission of IEEE.

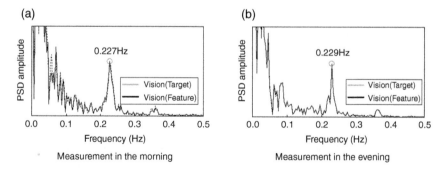

Figure 3.45 Power spectral distribution: (a) measurement in the morning; (b) measurement in the evening. *Source:* Reproduced with permission of IEEE.

(a)

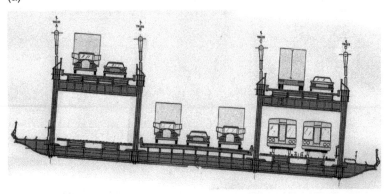

(b)

Figure 3.46 Manhattan Bridge: (a) cross-section; (b) test setup. *Source:* Reproduced with permission of Elsevier.

This test focuses on evaluating the long-distance, multipoint measurement capability of the computer vision sensor system.

The vision sensor system is placed on stable stone steps with negligible ground motion approximately 300 m from the bridge's mid-span point. For this flexible long-span bridge, a frame rate of 10 fps is used. The known dimensions (7.2 m, as shown in Figure 3.47a) of the vertical trusses are used to estimate the scaling factor.

First, a single-point displacement measurement is tested. The vertical displacement at a single point in the mid-span region, as shown in Figure 3.47a, is measured during the passage of subway trains. The scaling factor is 20.5 mm/pixel. The dynamic displacement response recorded for a total of 800 seconds is plotted in Figure 3.48. The bridge response displacement to the subway trains can be clearly distinguished, reaching a peak amplitude of 356 mm. This value is similar to those measured by both the GPS and interferometric radar systems in literature [39].

(a)

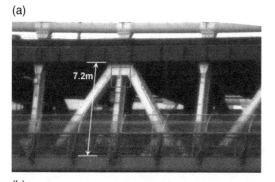

(b)

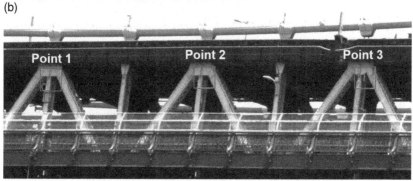

Figure 3.47 Tracking target on the bridge: (a) one target; (b) simultaneous measurements of three targets. *Source:* Reproduced with permission of Elsevier.

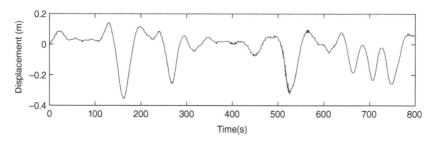

Figure 3.48 Displacement measurement of one target. *Source:* Reproduced with permission of Elsevier.

Second, simultaneous multipoint displacement measurements are tested. The camera lens is zoomed out to obtain a large field of view (FOV), i.e. area that is visible in the image, and three measurement points at the mid-span region are selected, as shown in Figure 3.47b. The vertical displacements at these three points are simultaneously measured by the single camera, and they are plotted in

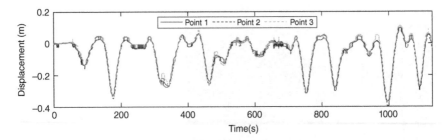

Figure 3.49 Simultaneous displacement measurements of three targets. *Source:* Reproduced with permission of Elsevier.

Figure 3.49. Compared with the single-point measurement in Figure 3.48, the three-point measurements in Figure 3.49 display more fluctuations, especially when the displacement amplitude is small. This is because the scaling factor is increased to 36 mm/pixel as the FOV is broadened to include the three measurement points, resulting in decreased measurement resolution compared to the single-point measurement case. When using the vision sensor to simultaneously measure multiple points on the structure, a trade-off needs to be made between the size of the FOV and the measurement resolution.

When simultaneously measuring the full-field displacements of a large-scale structure such as a high-rise building or a long-span bridge, multiple synchronized cameras can be used to target different sections of the structure, which helps to reduce the FOV of each camera and increase the measurement resolution. A method of synchronization is described in Figure 2.3.

The field test conducted on the Manhattan Bridge further investigates the camera vibration problem and a practical solution based on simultaneous multipoint measurement. As shown in Figure 3.50, the FOV of the camera is further broadened to include a building in the background of the suspension bridge. The vertical displacements of the building and the bridge at the two targets (framed by squares in the picture) are simultaneously measured. The camera vibration can be estimated given a reasonable assumption that the building is not moving in the vertical direction, and thus the measured building displacement is actually the camera vibration. Figure 3.51 plots the displacements of both the camera and the bridge mid-span (in pixels) during the passage of a train. It is concluded that compared to the bridge displacement, the camera motion is negligible in this field test.

This practical solution to the camera vibration problem is limited, however. First, in the field, it is not always easy to find a stationary reference point in the FOV of the image. Second, when enlarging the FOV to include such a reference point, the resolution at each measurement point is reduced. Although sub-pixel techniques, discussed earlier, can help improve resolution, they are still limited by the camera hardware. Trade-offs between measurement resolution and FOV are necessary.

Figure 3.50 Tracking targets on the bridge and the background building.
Source: Reproduced with permission of Elsevier.

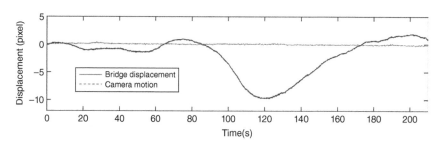

Figure 3.51 The camera motion and the mid-span vertical displacement of the bridge.
Source: Reproduced with permission of Elsevier.

3.10 Summary

This chapter described a wide range of laboratory and field experiments carried out in order to evaluate the robustness and accuracy of the computer vision sensor system for dynamic displacement measurements. The measurement results on laboratory structural models and field bridges given a variety of excitations, including sinusoidal, random, seismic, and moving vehicles and trainloads, were compared with those using conventional on-structure sensors. The two represent-ative template-matching algorithms incorporated in the computer vision system, UCC and OCM, were tested and compared in both ideal and unfavorable environmental conditions. The performance of the vision sensor tracking low-contrast natural markers on a structural surface was compared with that using high-contrast artificial markers. Simultaneous measurement of multiple points using a single camera were also studied. The practical method for obtaining the scaling factor was also investigated. Major findings and conclusions are summarized as follows.

1) While the image intensity-based UCC and gradient-based OCM template-matching algorithms both perform well in ideal environments, the OCM algorithm exhibits excellent robustness against challenging conditions such as fluctuating illumination, insufficient illumination, partial template occlusion, and background disturbances, which are frequently encountered in the field. Therefore, a computer vision sensor incorporating a gradient-based template-matching algorithm such as OCM will be more reliable than an intensity-based algorithm for use in challenging outdoor environments for field measurements and long-term monitoring of structures.

2) The computer vision sensor can achieve high measurement accuracy by tracking, from hundreds of meters away, natural markers on the structural surface such as rivets and corners. This can be done even with dim illumination at night. The elimination of the need to access the structure to install high-contrast artificial markers makes the computer vision sensor truly cost-effective and easy to operate, compared with conventional on-structure sensors. This also makes it possible to choose any points to measure displacements by post-processing a signal video recording, which is another unique and significant advantage of the vision sensor.

3) Enhanced by subpixel techniques, the computer vision system is capable of simultaneous, accurate measurements of structural displacements at multiple points using a single camera. Trade-offs need to be made between the size of the FOV and the measured displacement resolution.

4) A practical method is recommended for estimating the scaling factor based on a known physical dimension on the object surface and the corresponding image dimensions in pixels. For cases with a non-perpendicular lens optical axis, scaling factors in both the horizontal and vertical directions should be estimated. Satisfactory measurement accuracy can be achieved even with small camera tilt angles.

4

Application in Modal Analysis, Model Updating, and Damage Detection

Experimental modal analysis has developed into a major technology for the study of structural dynamics. Complex structural dynamic responses can be represented using decoupled vibration modes, each of which is a single-degree-of-freedom (DOF) system governed by its natural frequency and mode shape. The study of the dynamic characteristics represented by modal properties, including the natural frequency and mode shape, is referred to as *modal analysis*. Theoretical modal analysis solves the differential equations of the motion (i.e. the analytical model) of a structure to compute the structure's modal properties. Experimental modal analysis, on the other hand, primarily concerns the determination of the modal properties by conducting vibration tests on the structure and analyzing the acquired input and output measurement data. Whether the object is a turbine blade rotating at high speed or a bridge sustaining traffic and strong wind loads, experimental modal analysis has been widely applied to a variety of mechanical, aerospace, and power-generation structures, primarily to validate and improve their structural designs.

Modal properties identified from an experimental modal analysis can be further used to determine the structural parameter values of a finite element model of the structure by matching experimentally identified modal properties with computed ones. Such parameters of the analytical model can be periodically updated through experimental modal analysis. This process is referred to as *model updating*. Such an updated analytical model can be used to reliably predict structural behavior under given loads. It can also be used for damage-detection purposes, since damage such as cracks causes a local reduction in stiffness and an increase in damping [40, 41]. By analyzing changes in, for example, stiffness at the finite element level, structural damage can be not only detected but also located. This approach, based on

Computer Vision for Structural Dynamics and Health Monitoring, First Edition.
Dongming Feng and Maria Q. Feng.
© 2021 John Wiley & Sons Ltd.
This Work is a co-publication between John Wiley & Sons Ltd and ASME Press.
Companion website: www.wiley.com/go/feng/structuralhealthmonitoring

(a)

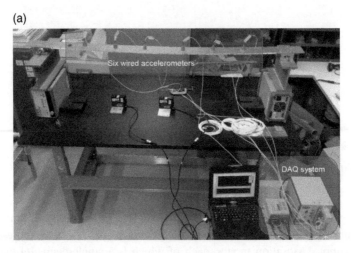

(b)

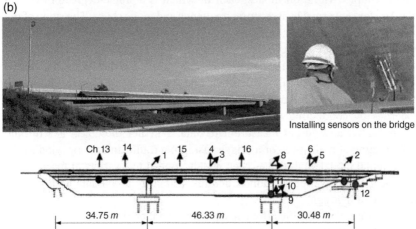

Figure 4.1 Typical modal testing and SHM systems using accelerometers: (a) modal testing of a beam in the lab; (b) long-term SHM of the Jamboree bridge.

vibration testing and model updating, has recently been studied for long-term structural health monitoring (SHM) and tested on highway bridges [42, 43].

Figure 4.1 shows two representative setups: one for experimental modal testing of a laboratory model structure (a beam) and the other for long-term SHM of a highway bridge, the Jamboree Road Overcrossing in Irvine, CA. Typically, an array of accelerometers is mounted on the structure to measure the structural responses to excitations, such as impact force on the lab model structure and vehicle loading on the bridge. One main issue, particularly for a large-size structure in the field, is the high cost of installing and maintaining the accelerometers.

Practical application of SHM technology in actual structures is, to a large extent, limited by the requirement of installing and maintaining the sensor systems. Compared with conventional contact-type on-structure sensors such as accelerometers, the noncontact computer vision sensor is far more cost-effective and agile in implementation, as confirmed by the various tests described in Chapter 3. In addition, the computer vision sensor provides a unique advantage that any point on the structure can be selected to measure its displacement from a signal video recording.

Although the emerging computer vision sensor technology has demonstrated its high accuracy and robustness for multipoint structural displacement measurements (as presented in Chapter 3), study of its application for SHM is still at an early stage. This chapter demonstrates the usefulness of structural displacement data for experimental modal analysis, finite element model updating, and damage detection. Specifically, Section 4.1 demonstrates the application of the computer vision sensor data for experimental modal analysis of the beam and the three-story frame structure presented in Chapter 3; Section 4.2 presents finite element model updating of the frame structure based on identified modal properties; and Section 4.3 discusses a method and its experimental validation for damage detection using a mode shape curvature (MSC)-based damage index obtained using dense vision sensor measurements.

4.1 Experimental Modal Analysis

Identifying modal parameters based on structural vibration tests is the heart of experimental modal analysis, and various testing and data analytics techniques have been developed over the past several decades – which are beyond the scope of this book. Rather, this section focuses on how the computer vision sensor can be successfully used for experimental modal analysis, in comparison with conventional sensors, through vibration testing of two laboratory-scale model structures.

4.1.1 Modal Analysis of a Frame

The first modal test is conducted on the scaled three-story frame structure shown in Figure 3.4 in Chapter 3. The structure is fixed on a shaking table and subjected to white noise excitations in the horizontal direction. Modal properties of the frame, including natural frequencies and mode shapes, are identified from the measured input excitation and output structural response time histories.

The aluminum frame structure is bolt-connected for all column-floor connections. The structural material and geometry parameters are summarized

Table 4.1 Parameters of the three-story frame structure.

Column		Floor	
Parameters	**Value**	**Parameters**	**Value**
Mass density (kg m³)	2700	Mass density (kg m³)	2700
Elastic modulus (GPa)	69	Elastic modulus (GPa)	69
Cross-section (mm²)	2.25×25	Area (mm²)	240×240
Length (mm)	200	Thickness (mm)	16.5

Source: Reproduced with permission of John Wiley & Sons.

in Table 4.1. Using these parameter values and assuming fixed column-floor connections, a lumped mass-spring analytical model is created.

This same setup, as presented in Figure 3.4, Section 3.2, was previously used for the experimental evaluation of the computer vision sensor's performance for tracking low-contrast natural targets (four bolt connections) compared with tracking four high-contrast artificial targets, as well as four high-accuracy reference sensors. The CMOS camera described in Table 2.1 is used. The camera captures video images at a sampling rate of 150 fps. The reference laser displacement sensors (LDSs) are installed between each floor of the frame model at stationary reference points. Additionally, four accelerometers are installed on each floor to further compare the experimental modal analysis results.

During the white noise excitations of the shaking table, both the table excitation at the frame base and the frame floor response time histories are measured by the computer vision sensors tracking the natural and artificial targets, respectively, and the reference LDSs and accelerometers. Segments of the measured base (i.e. table) and floor displacement time histories are shown in Figures 3.6 and 3.7.

Modal properties, including the natural frequencies and mode shapes of the frame, are identified using the Eigensystem Realization Algorithm (ERA). Table 4.2 compares the first three computed natural frequencies based on the design parameters of the frame (i.e. the initial values) with those identified from the displacements measured by the LDSs, the OCM-based vision sensor (which tracks the artificial targets), and the OCM-based vision sensor (which tracks the natural targets), and from the acceleration measurements by the accelerometers. The identified first frequency ranges from 6.52 to 6.56 Hz, exhibiting very small discrepancies among the results from the four different sensor measurements. The same observation can be made regarding the second and third frequencies.

The frequencies of the frame identified from modal testing are larger than those computed using the analytical model. This is mainly due to the much larger lateral

Table 4.2 Comparison of identified natural frequencies of the frame structure.

| Natural freq. (Hz) | Computed from the initial model | OCM-based vision sensor | | LDS | Accelerometer |
		Artificial target	Natural target		
First	5.68	6.52	6.56	6.55	6.52
Second	15.82	19.44	19.56	19.49	19.35
Third	22.66	28.08	28.18	28.10	27.98

Source: Reproduced with permission of John Wiley & Sons.

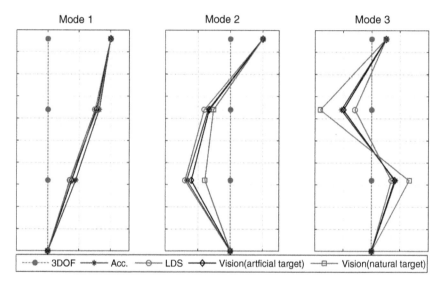

Figure 4.2 Comparison of identified mode shapes of the frame structure.

stiffness of columns (or smaller effective column length) resulting from the rigid column-to-floor angle connections. This discrepancy demonstrates the need to update an initial analytical model derived from its design parameters. In fact, a constructed structure never agrees perfectly with its design, and experimental modal analysis of the structure can be conducted to update the initial analytical model based on the design parameters.

Figure 4.2 compares the first three mode shapes identified from the acceleration measurements by the accelerometers with the displacement measurements from the LDS, vision sensor (targeting the artificial targets), and vision sensor (targeting the natural targets). They all match well, particularly the first mode shapes.

Note that the mode shapes are scaled to 1 with respect to a reference point (here, the third floor).

In summary, the modal properties of the frame structure identified by the computer vision sensor measurements agree well with those identified by the conventional laser displacement sensors and the accelerometers. However, compared with the accelerometers, the vision sensor is more convenient and cost-effective to set up for experimental modal analysis, because displacements of the structure at multiple points can be measured from one camera in a noncontact fashion by tracking existing natural markers on the structural surface.

Note that there are many important signal-processing aspects of experimental modal analysis, which are beyond the scope of this book. For this frame structure subjected to white noise excitations, averaging and windowing are applied to the measured data to reduce the random noise and alleviate the leakage phenomenon, respectively. Here, the MATLAB function *cpsd* performs the two steps. For example, (P_{xy}, F) = cpsd (x, y, window, noverlap, nfft, fs) estimates the cross-power spectral density P_{xy} of the discrete-time signals x and y using the Welch's averaged, modified periodogram method of spectral estimation. By default, signals x and y are divided into the longest possible segments to obtain close to but not exceed 8 segments with 50% overlap. Each segment is windowed with a Hamming window. The modified periodograms are averaged to obtain the PSD estimate.

The following are the MATLAB commands for using the ERA for modal analysis and mode shape plots.

MATLAB Code – Modal Analysis Using ERA

```
clear;close all;clc
%--------------- Load input and output data
load ground.csv        % 1st column:time; 2nd
                         column:displacement input on
                         base floor
load first_floor.csv   % 1st column:time; 2nd
                         column:displacement output of
                         1st floor
load second_floor.csv  % 1st column:time; 2nd
                         column:displacement output of
                         2nd floor
load third_floor.csv   % 1st column:time; 2nd
                         column:displacement output of
                         3rd floor
```

```
input=ground(:,2);     % input:  n by 1
Disps=[ground(:,2) first_floor(:,2) second_floor(:,2)
third_floor(:,2)];     % output: n by 4


fs = 150;              % sampling frequency
dt = 1/fs;             % time interval
n=length(input);       % input data length
t=0:dt:(n-1)*dt;       % time


Nout = 4;              % number of output
Ninput = 1;            % number of input
Nfft = 4096;           % FFT number
df = 1/(dt*Nfft);      % Frequency resolution

%--------------- Conduct ERA for modal analysis
[csdxx,Fxx]= cpsd(input,input,[],Nfft/2,Nfft,fs);
% PSD of the input signal.
for ii = 1:Nout
    % Estimate the cross power spectral density of
      input and output signals
    [csdxy(:,ii), Fxy]= cpsd(input,Disps(:,ii),[],Nfft
    /2,Nfft,fs);
    % Obtain the transfer function from H = Sxy/Sxx.
    Hxy(:,ii) = csdxy(:,ii)./csdxx;
    % Mirroring with complex conjugate
    Htemp(:,ii)=[Hxy(:,ii); conj(flipud(Hxy
    (1:end-1,ii)))];
end


% Produce Impulse response function from the mirrored
transfer function Htemp
for ii = 1:Nout
    % Inverse FFT (frequency domain to time domain)
    YY(:,ii) = 1/dt * real(ifft(Htemp(:,ii), Nfft));
end
for ii = 1:Nout
    Y(ii,1,:)=YY(:,ii)';
end
```

```matlab
ncols=100; nrows=500;   % The number of columns and
                        rows for the Hankel matrix
cut=10;                 % Estimate the Singular Value
                        Cutoff, or mode numbers to be
                        identified
[fd,zeta,shapes,lambda,partfac,EMAC,sv,Aout,Bout,Cout]
=era(Y,fs,ncols,nrows,Ninput,cut);

%--------------- Plot first 3 mode shapes
Index=find(EMAC>=95);   % Extended modal amplitude
                        coherence (MAC)
index=Index(1:3);       % First 3 mode shapes
shape=[0 imag(shapes(2,index(1)))
imag(shapes(3,index(1))) imag(shapes(4,index(1))) ;
          0 imag(shapes(2,index(2))) imag(shapes
          (3,index(2))) imag(shapes(4,index(2))) ;
          0 imag(shapes(2,index(3))) imag(shapes
          (3,index(3))) imag(shapes(4,index(3)))];
y=[0 1 2 3];
x=[0 0 0 0];
figure(1)
plot(x+shape(1,:)/shape(1,4),y,'r-*');
% Plot 1st mode shape
hold on
plot(x,y,'b-*');
title(['1st mode shape:' num2str(fd(index(1)))
'Hz'],'FontSize',12)
xlim([-2 2])
set(gcf, 'Position',[200,200,250,400], 'color','w')
figure(2)
plot(x+shape(2,:)/shape(2,4),y,'r-*');
% Plot 2nd mode shape
hold on
plot(x,y,'b-*');
title(['2nd mode shape:' num2str(fd(index(2)))
'Hz'],'FontSize',12)
xlim([-2 2])
set(gcf, 'Position',[200,200,250,400], 'color','w')
figure(3)
plot(x+shape(3,:)/shape(3,4),y,'r-*');
% Plot 3rd mode shape
```

```
hold on
plot(x,y,'b-*');
title(['3rd mode shape:' num2str(fd(index(3)))
'Hz'],'FontSize',12)
xlim([-3 3])
set(gcf, 'Position',[200,200,250,400], 'color','w')
```

Comments:
Function [fd,zeta,shapes,lambda,partfac,EMAC,sv,Aout,Bout,Cout]=era(Y,fs,nco
ls,nrows,Ninput,cut) is the eigensystem realization algorithm for modal analysis. Please see book "Identification and Control of Mechanical Systems" by
Juang, J. N. and Phan, M. Q., Cambridge University Press, 2001.

% Function input data
% Y is the impulse response function, in form of Y(ouputs,inputs,data sequence)
% fs: sampling frequency in Hz
% ncols is the no. of columns in the Hankel matrix;
% nrows is the no. of rows in the Hankel matrix
% cut: estimate the singular value cutoff, or mode numbers to be identified
% Function output data
% fd: identified undamped frequencies;
% zeta: identified damping ratios of each mode shape
% shapes: mode shapes
% partfac: modal participation factors

4.1.2 Modal Analysis of a Beam

The second example of modal testing is on the scaled beam structure shown in
Figure 3.16 in Chapter 3. Hammer impact tests are conducted on the beam to
identify its modal properties, including natural frequencies and mode shapes. The
structural material and geometrical parameters are summarized in Table 4.3.

Table 4.3 Parameters of the simply supported beam.

Parameters	Value
Mass density	$2700\,\mathrm{kg\,m^3}$
Elastic modulus	$69\,\mathrm{GPa}$
Cross-section	$101.6 \times 4.76\,\mathrm{mm^2}$
Length	$1.6\,\mathrm{m}$

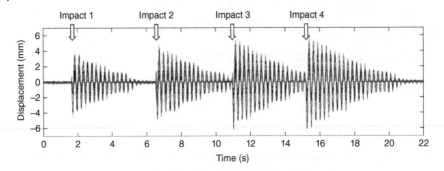

Figure 4.3 Displacement measurements at points 2–31 by the vision sensor.

Using the structural parameter values and assuming simple supports, an initial analytical model of the beam is created.

The computer vision sensor system uses one camera to measure the beam's vertical displacements at 30 points simultaneously by tracking the 30 dots marked on the side surface of the beam, as shown in Figure 3.17. The camera is placed 10 m from the beam, with the optical axis perpendicular to the side surface. The scaling factor is 1.20 mm/pixel. For comparison purposes, two LDSs are installed to measure the beam displacement at points 9 and 16, and six accelerometers are attached to the beam at points 6, 10, 14, 19, 23, and 27. All three types of sensors use the same data-acquisition rate of 50 Hz.

Here, the hammer impact test is conducted in order to excite and measure a broader range of beam vibration frequencies. The hammer impact is applied at point 4, and the displacement time histories simultaneously measured at the 30 points along the beam are shown in Figure 4.3. The displacement time histories measured at points 9 and 16 by the vision sensor are compared with those from the LDSs. As shown in Figure 4.4, excellent agreement is observed between the two sensors, with normalized root mean squared errors (NRMSEs) of 1.8 and 1.2%, respectively, for the two measurement points 9 and 16.

Multiple impact tests are conducted, and beam response time histories are measured, from which averaged PSD functions are obtained. Figure 4.5 compares the corresponding PSD amplitudes measured by the vision sensor at 30 points, the LDSs at two points, and the accelerometers at six points. The results of the three types of sensors show a perfect match. They all clearly identified the first mode frequency of 4 Hz and the second mode at 15.82 Hz. Since a 50 Hz sampling frequency is used for all three types of sensors, only the first two natural frequencies can be obtained.

Subsequently, the mode shapes of the simply supported beam model can be obtained through modal analysis. Figure 4.6 compares the first two normalized mode shapes measured by the accelerometers and the vision sensor. They agree

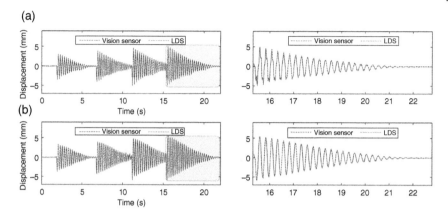

Figure 4.4 Comparison of displacement measurements (a) at point 9; (b) at point 16.

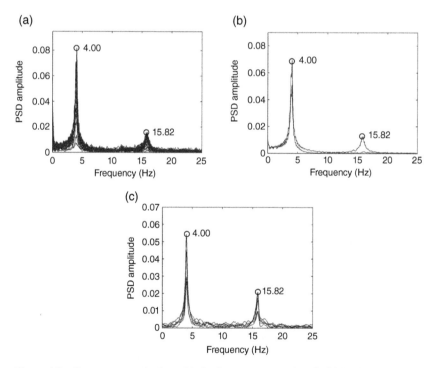

Figure 4.5 Frequency results from (a) displacements at points 2–31 by the vision sensor; (b) displacements at points 9 and 16 by LDS; (c) accelerations at six points by accelerometers. *Source:* Reproduced with permission of Elsevier.

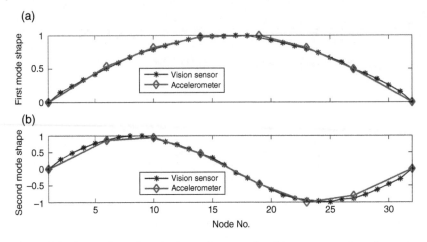

Figure 4.6 Comparison of mode shapes between the vision sensor and accelerometer:
(a) first mode shape; (b) second mode shape. *Source:* Reproduced with permission of Elsevier.

Table 4.4 Comparison of identified natural frequencies of the beam structure.

Frequency (Hz)	Computed/baseline	Modal testing	
		Free vibration	Hammer impact
First	4.53	4.0	4.0
Second	18.13	15.82	15.82

very well. But the vision sensor results in smooth mode shapes due to the densely distributed sensor points, while the spatial resolution of the mode shapes from the acceleration data is limited by the number of accelerometers.

Table 4.4 compares the frequencies of the beam computed by the initial analytical model with those identified from the impact tests. The identified frequencies are smaller than the computed ones, primarily because the initial analytical model does not include the additional masses of the accelerometers attached to the beam. This is another drawback of conventional on-structure sensors, whose masses may affect the accuracy of the structural modal parameter identification results. It is especially problematic for small-scale, lightweight structures or scaled testing models of large structures. In contrast, the noncontact vision sensor does not have such problems, which is another advantage.

Furthermore, compared with the accelerometers and the LDSs, the vision sensor system is far more convenient and cost-efficient for experimental modal analysis, since a single camera can simultaneously measure displacements of a large number of points in a noncontact fashion, and setting up the camera takes little time.

4.2 Model Updating as a Frequency-Domain Optimization Problem

An accurate analytical model of a structure that reflects the current structural condition is highly desired for detecting structural damage, predicting structural behavior given new loading scenarios, analyzing remaining life, and improving the future structural design. Such a model can be developed by using design drawings and documentation as an initial baseline. As discussed earlier, this initial baseline model should be updated by fine-tuning the values of its structural parameters (such as stiffness) using in situ vibration measurement data, to better represent the current structural condition. The discrepancy between designed and identified structural parameters from the measurements can be due to uncertainties such as material properties, incomplete design information, and construction errors, as well as long-term material deterioration, changes of geometries and boundary conditions, etc. Updating the baseline model based on continuous in situ measurements will enable the timely detection of changes in structural parameters and assessment of structural damage. The existing model-updating methods in the literature are generally based on experimental modal analysis. For example, Feng et al. [9] and Soyoz and Feng [43] presented a finite element model-updating method based on in situ ambient vibration tests and modal analysis and applied it to update a highway bridge model and monitor long-term structural health.

The model-updating process in the frequency domain can be considered an optimization problem, the objective of which is to find the best parameter estimate $\hat{\theta}$ for the unknown structural parameters $\theta = \left\{ \theta_1, \theta_2, \cdots, \theta_{n_\theta} \right\}^T$ by minimizing discrepancies between analytical and experimental modal data, including natural frequencies and mode shapes of the dominant modes [31]. Here, the objective function can be defined as the sum of weighted least-square errors:

$$\Pi(\theta) = \mathbf{r}(\theta)^T \mathbf{W} \mathbf{r}(\theta) \tag{4.1}$$

where $\mathbf{r}(\theta)$ is the residue vector and $\mathbf{W}$ is the weighting matrix, given by

$$\mathbf{r}(\theta) = \begin{bmatrix} \mathbf{r}^f(\theta) \\ \mathbf{r}^s(\theta) \end{bmatrix}, \ \mathbf{W} = diag\left(\ldots, \omega^{f_i}, \ldots, \omega^{\phi_i}, \ldots\right)$$

with

$$\mathbf{r}^f(\theta) = \begin{bmatrix} \dfrac{f_i^{FEM}(\theta) - f_i^M}{f_i^M} \end{bmatrix} \tag{4.2}$$

$$\mathbf{r}^s(\theta) = \begin{bmatrix} \dfrac{\phi_i^{l,FEM}(\theta)}{\phi_i^{r,FEM}(\theta)} - \dfrac{\phi_i^{l,M}}{\phi_i^{r,M}} \end{bmatrix}, (l \neq r), i \in \{1, 2, \ldots, n_m\}$$

where $\mathbf{r}^f(\theta)$ and $\mathbf{r}^s(\theta)$ denote the frequency and mode shape residue vectors, respectively; f_i^{FEM} and ϕ_i^{FEM} denote the ith numerical frequency and mode shape vector, respectively; f_i^M and ϕ_i^M refer to the ith experimental frequency and mode shape vector, respectively; ω^{f_i} and ω^{ϕ_i} are the weighting factors for the ith natural frequency and mode shape, respectively; and n_m is the number of identified frequencies and mode shapes. In $\mathbf{r}^s$, the numerical and experimental mode shapes are scaled to 1, with the indices l and r in ϕ_i^{FEM} and ϕ_i^M denoting an arbitrary and a reference DOF in the ith mode shape. "$l \neq r$" means the reference DOF is not counted in the residue vector. Note that in $\mathbf{r}^f$, relative differences are taken in order to have a similar weight for each frequency.

As an example, this approach is applied to the three-story frame structure to update its analytical model based on the experimental modal analysis, as presented in Section 4.1.1. Here, the first three natural frequencies and mode shapes identified in the previous section are used to update the three inter-story stiffness $\theta = \{k_1, k_2, k_3\}^T$ of the frame structure. Therefore, the residue vector $\mathbf{r}$ contains a total of nine residuals. Weighting factors for the residues should be selected based on their importance and measurement accuracy. In most cases, modal frequencies can be identified from measurements more reliably than mode shapes and thus should be given more weight. Here, a weighting factor of 1 is adopted for all frequency residues and 0.1 for all mode shape residues. In the optimization process, the initial parameter values are set to $\tilde{k}_1 = \tilde{k}_2 = \tilde{k}_3 \approx 1.8 \times 10^4 \, N/m$, which are the initial values computed based on the design parameters of the frame structure. The optimization problem is solved using the sensitivity-based damped Gauss-Newton optimization algorithm, described as follows:

Step 1: Choose initial parameter values $\theta^{(0)}$, select the non-negative numerical damping factor $h^{(0)}$, and set the iteration number index $s = 0$.

Step 2: Calculate the Jacobian matrix of the residual function using the central difference method:

$$\left[\mathbf{J_r} \right]_{ij} = \frac{\partial \mathbf{r}_i\left(\theta^{(s)}\right)}{\partial \theta_j} \tag{4.3}$$

Step 3: Update the parameter set:

$$\theta^{(s+1)} = \theta^{(s)} + \left[\mathbf{J_r}^T \mathbf{J_r} + h^{(s)} \mathrm{diag}\left(\mathbf{J_r}^T \mathbf{J_r} \right) \right]^{-1} \mathbf{J_r}^T \mathbf{r}_i\left(\theta^{(s)}\right) \tag{4.4}$$

Step 4: Update the numerical damping factor $h^{(s)}$ using the adaptive tracking approach.

Step 5: Update the iteration index $s = s + 1$.

Figure 4.7 Stiffness optimization evolution using measurements taken by the vision sensor (natural target). *Source:* Reproduced with permission of John Wiley & Sons.

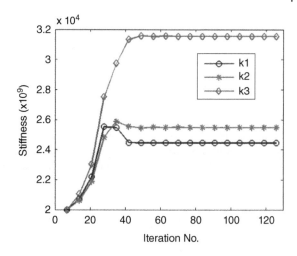

Table 4.5 Stiffness identification results ($\times 10^4 N/m$).

	Initial baseline stiffness	OCM-based vision sensor		Accelerometer
		Artificial target	Natural target	
First floor: k_1	1.80	2.40	2.44	2.42
Second floor: k_2	1.80	2.51	2.54	2.54
Third floor: k_3	1.80	3.13	3.15	3.09

Source: Reproduced with permission of John Wiley & Sons.

Step 6: Repeat Steps 2–5 until the convergence criterion is met, e.g. the relative error of the objective function is smaller than a tolerance value (e.g. 1×10^{-6}).

For the example three-story frame structure, the natural frequencies and mode shapes are identified from the experimental modal analysis using the displacements measured by the computer vision sensors and the reference sensors, as presented in Section 4.1.1. For illustration purposes, Figure 4.7 depicts the evolution process of the inter-story stiffness identification, which uses the modal properties extracted from the displacement measurements taken by the vision sensor targeting the natural targets. It is observed that all three stiffness quantities converge quickly.

Table 4.5 shows the identified stiffnesses k_1, k_2, and k_3 from the modal properties measured by the vision sensor (artificial targets), vision sensor (natural targets), and accelerometers. The differences in the identified stiffness values between the vision displacement sensors (tracking natural and artificial targets) and the accelerometers are very small.

Table 4.5 also lists the baseline stiffness values computed from the initial lumped mass-spring model of the frame structure using the design parameters and the assumption of fixed joints in the floor beams and columns. The identified stiffness values are larger than the initial baseline values, which is caused by the reduction of the effective column length due to the bolted column-floor angle connections. This discrepancy demonstrates the need to update an analytical model derived from design parameters. Experimental modal analysis, combined with the frequency-domain optimization procedure, provides a powerful tool for updating structural parameters such as stiffness. For a real structure in the field, such as a bridge, for which input excitation is difficult to measure, model updating can be carried out using measured structural responses only while assuming input excitations as white noise [9, 43] or constructing a stochastic input excitation model based on partial vehicle information measured by computer vision [44, 45].

Updating the model of the three-story frame structure is carried out using the following MATLAB scripts.

MATLAB Code – Model Updating of the Three-Story Frame Structure

```
%-------------------------
% Description: Main script
%-------------------------
close all; clear all; clc;
tic
global shape_m fq_m m1 m2 m3
m1 = 2.906; m2 = 2.906; m3 = 2.706;

%---------- Read identified modal parameters
i=input('Select identified modal parameters from
measurements by Accelerometer(1), Vision-artificial
target(2), or Vision-natural target(3): ');

if i==1
% Read identified modal parameters from measurements
by Accelerometer
  Mode_shape=load('Acc_modeshape.mat');
  fq_m=[6.52 19.35 27.98];
elseif i==2
```

```
% Read identified modal parameters from measurements
by Vision-artificial target
  Mode_shape=load('Vision_modeshape_OCM.mat');
  fq_m=[6.52 19.44 28.08];
elseif i==3
% Read identified modal parameters from measurements
by Vision-natural target
  Mode_shape=load('Vision_modeshape_OCM_bolt');
  fq_m=[6.56 19.56 28.18];
end
shape_m=Mode_shape.shape;

%----------- Model updating using damped Gauss-Newton
method
global history2
history2.x = [];
history2.funccount = [];

% Set initial Conditions
ic = 1e4*[1.8 1.8 1.8];

% Set lower and upper bounds
lb = 0.5e4*[1 1 1];
ub = 4e4*[1 1 1];

% Set optimization options
opts = optimset('fmincon');
opts.Display = 'iter';
opts.Algorithm = 'active-set';
opts.MaxFunEvals = 1000;
hStep = 1e-4;
dXOpt = hStep*ones(1, 3);
Func = 'objectiveFun';
MCrun = 1;                    % Monte Carlo simulations
XOptMC = zeros(MCrun, 3);

for r = 1:MCrun
  Options = optimset('Algorithm', 'trust-region-
reflective', 'Display', 'iter-detailed', 'ScaleProb-
lem', 'Jacobian', 'TolFun', 1e-12, ...
```

```matlab
'FinDiffType', 'central', 'FinDiffRelStep', dXOpt,
'MaxIter', 100, 'TolX', 1e-10, 'UseParallel',
'always', 'OutputFcn', @outfun2);
  [XOpt, ~, ~, ~, ~, ~, ~] = lsqnonlin(Func, ic, lb, ub,
Options);
  XOptMC(r, :) = XOpt;
end


%----------- Plot stiffness optimization evolution
figure (1)
plot(history2.funccount,history2.x(:,1),'b-o',
'linewidth', 1.0)
hold on
plot(history2.funccount,history2.x(:,2),'m-*',
'linewidth', 1.2);
plot(history2.funccount,history2.x(:,3),'r-d',
'linewidth', 1.2);
legend('k1','k2','k3')
hh = legend('k1','k2','k3');
set(hh, 'fontsize', 11, 'Location','Northwest');
legend boxoff
xlabel('Iteration No.', 'fontsize', 12)
ylabel('Stiffness (\times10^9)','Fontname','Arial',
'fontsize', 12)
set (gcf, 'Position', [200,200,400,300], 'color', 'w');


%-------------------------------
% Description: Objective function
%-------------------------------
function ObjVal = objectiveFun(k)

global shape_m fq_m m1 m2 m3

k1 = k(1); k2 = k(2); k3 = k(3);
M = diag([m1 m2 m3]);
K = [k1+k2 -k2 0; -k2 k2+k3 -k3; 0 -k3 k3];

[V,D]=eig(K,M);
D=diag(D);
fq=sqrt(D)/(2*pi)';
shape_FEM=[[0;0;0],V'];
```

```
for i=1:3
    shape_m(i,:)=shape_m(i,:)/shape_m(i,4);
    shape_FEM(i,:)=shape_FEM(i,:)/shape_FEM(i,4);
end

w=[1,0.8,0.5];
alpha=1;
beta=0.1;

%% complete information: all modal information
R=[fq_m(1)-fq(1);                    % Residue vector
    fq_m(2)-fq(2);
    fq_m(3)-fq(3);
    shape_m(1,2)-shape_FEM(1,2);
    shape_m(1,3)-shape_FEM(1,3);
    shape_m(2,2)-shape_FEM(2,2);
    shape_m(2,3)-shape_FEM(2,3);
    shape_m(3,2)-shape_FEM(3,2);
    shape_m(3,3)-shape_FEM(3,3)];

W=[alpha*w(1) 0 0 0 0 0 0 0 0;        % Weighting matrix
    0 alpha*w(2) 0 0 0 0 0 0 0;
    0 0 alpha*w(3) 0 0 0 0 0 0;
    0 0 0 beta*w(1) 0 0 0 0 0;
    0 0 0 0 beta*w(1) 0 0 0 0;
    0 0 0 0 0 beta*w(2) 0 0 0;
    0 0 0 0 0 0 beta*w(2) 0 0;
    0 0 0 0 0 0 0 beta*w(3) 0;
    0 0 0 0 0 0 0 0 beta*w(3)];
ObjVal = sqrt(W')*R;ObjVal = sqrt(W')*R;


%-----------------------------------------
% Description: Store the iterative results
%-----------------------------------------
function stop = outfun2(x, optimValues, state)
global history2
stop = false;
switch state
case 'iter'
        history2.x = [history2.x; x];
```

```
            history2.funccount = [history2.funccount;
            optimValues.funccount];
end
end
```

4.3 Damage Detection

From periodic model updating, it is possible to establish a long-term trend of changes in structural parameters due to structural aging and deterioration and to detect sudden damage caused by unusual events such as vehicle collisions and earthquakes. Many approaches have been developed for detecting, locating, and assessing structural damage based on vibration testing/measurement and model updating [40]. This section, however, focuses on a simpler damage-detection method based on experimental modal analysis without involving model updating. This MSC-based method is enabled by the unique advantage of the computer vision sensor for dense (i.e. high spatial resolution) measurements.

To demonstrate this method, an experiment is carried out on the simply supported beam structure (described earlier in Section 4.1.2) in which damage is introduced. The OCM-based subpixel-resolution vision sensor with a single camera is used to measure beam response displacements at the 30 target points simultaneously under hammer impacts. Model shape indexes are constructed for both the damaged and undamaged structures to detect and locate the damage.

4.3.1 Mode Shape Curvature-Based Damage Index

MSC has been widely recognized as a damage-sensitive index to detect local structural damage. Pandey et al. [8] proposed the MSC-based method based on the premise that a reduction in structural stiffness associated with damage will cause an increase in the MSC, which can be calculated using the second-order central difference method as

$$\varphi_{q,j}'' = \frac{\varphi_{q-1,j} - 2\varphi_{q,j} + \varphi_{q+1,j}}{h^2} \tag{4.5}$$

where h is the distance between the measurement coordinates and $\varphi_{q,j}$ defines the modal displacement for the jth mode shape at the measurement coordinate q.

Damage information can be revealed by the peaks in the MSC changes between the damaged and undamaged structure, as given by the MSC-based damage index:

$$MSC_q = \sum_{j=1}^{N} \Delta\varphi_{q,j}'' = \sum_{j=1}^{N} \left| \varphi_{q,j}''^{,damaged} - \varphi_{q,j}''^{,intact} \right| \tag{4.6}$$

The sensitivity of this method to structural damage obviously depends on the spatial resolution of the mode shape index constructed based on measurements from modal testing. The practical application of this method has been limited due to the requirement of dense sensor instrumentation, but this limitation can be overcome with the computer vision sensor.

4.3.2 Test Description

As shown in Figure 4.8a, the simply supported beam model is made from rectangular aluminum sheet metal. The structural material and geometry parameters are summarized in Table 4.3. For the damaged beam model, two saw-cut dents between points 16 and 17 are introduced to simulate a 20% stiffness reduction, as illustrated in Figure 4.8c. Thirty black dots, numbered 2–31, are attached along the beam as targets for motion tracking.

During the hammer impact tests, 1280 × 240 pixel, 8-bit grayscale video images of the beam are streamed into the computer through a USB 3.0 cable with a sampling rate of 50 fps. By processing the digital video images with subpixel OCM-based vision sensor software, displacement time histories at points 2–31 are simultaneously obtained. Figure 4.9 shows the time histories measured at the 30 points along the intact beam under repeated hammer impacts at point 4 of the intact beam.

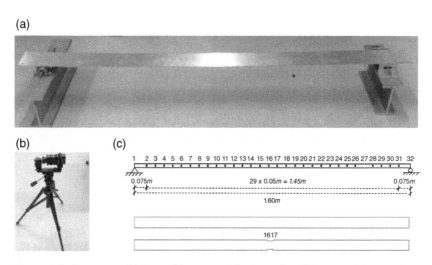

Figure 4.8 Test setup: (a) beam; (b) camera; (c) schematics of intact and damaged beams. *Source:* Reproduced with permission of John Wiley & Sons.

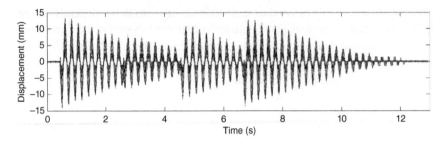

Figure 4.9 Displacement measurements at points 2–31. *Source:* Reproduced with permission of John Wiley & Sons.

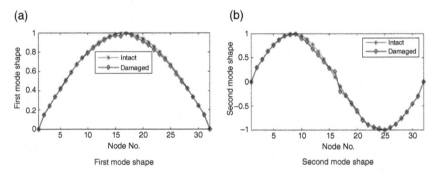

Figure 4.10 Identified first two mode shapes of the intact and damaged beams: (a) first mode shape; (b) second mode shape. *Source:* Reproduced with permission of John Wiley & Sons.

4.3.3 Damage Detection Results

The first two mode shapes of the undamaged and damaged beam structures, identified from the experimental modal analysis based on hammer impact testing, are plotted in Figure 4.10.

In Eq. (4.6), by setting $N = 2$, the MSC and modified mode shape curvature (MMSC) damage indices can be calculated based on the first two identified mode shapes as:

$$MSC_q = \sum_{j=1}^{2} \left| \varphi_{q,j}^{n,damaged} - \varphi_{q,j}^{n,intact} \right| \quad \text{and} \quad MMSC_q = \sum_{j=1}^{2} \left| \varphi_{q,j}^{n,damaged} - \varphi_{q,j}^{n,intact} \right|^2 \quad (4.7)$$

The MSC and MMSC damage indices for the damaged beam are obtained and plotted in Figure 4.11. The damage can be clearly identified and located from the peak in the MSC or MMSC index. The location of the peak perfectly matches the damage (dent) location. As shown in the figure, it would be impossible to use this method to detect and locate the damage without such a high spatial resolution

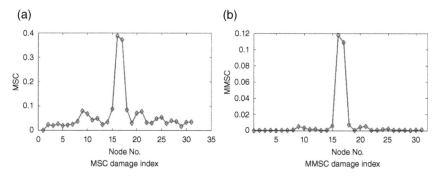

Figure 4.11 Damage indices of the damaged beam: (a) MSC damage index; (b) MMSC damage index. *Source:* Reproduced with permission of John Wiley & Sons.

MSC index enabled by the dense multipoint displacement measurement by the computer vision sensor. This dense spatial resolution measurement by a single camera is another unique advantage of the computer vision sensor over conventional pointwise sensors that need to be individually installed on the structure. For example, to achieve the same resolution using conventional accelerometers, 30 accelerometers would need to be installed on the beam.

MSC-based damage detection for the beam structure is achieved using the following MATLAB scripts.

MATLAB Code – MSC-Based Damage Detection for the Beam Structure

```
clear;clc;close all
load ModeShapeData.mat     % Include vectors "Model_
intact" and "Model_damaged"(size: 1x32)

%----------- 1st mode MSC
n=length(Model_damaged);
for j=2:n-1
  MSC1_intact(j)=(Model_intact(j+1)-2*Model_
  intact(j)+Model_intact(j-1))/h^2;
  MSC1_damaged(j)=(Model_damaged(j+1)-2*Model_
  damaged(j)+Model_damaged(j-1))/h^2;
end

%----------- 2nd mode MSC
n=length(Mode2_damaged);
```

```
for j=2:n-1
  MSC2_intact(j)=(Mode2_intact(j+1)-2*Mode2_
  intact(j)+Mode2_intact(j-1))/h^2;
  MSC2_damaged(j)=(Mode2_damaged(j+1)-2*Mode2_
  damaged(j)+Mode2_damaged(j-1))/h^2;
end

MSC_Index=abs(MSC2_intact-MSC2_damaged)+abs(MSC1_intact-
MSC1_damaged);
MMSC_Index=(MSC2_intact-MSC2_damaged).^2+(MSC1_intact-
MSC1_damaged).^2;

%----------- Plot 1st and 2nd mode MSC (modal shape
curvature)
figure(105)
subplot(1,2,1)
plot(MSC_Index,'rd','LineStyle','-','LineWidth',1.3,
'MarkerSize',5);
xlabel('Node No.', 'FontName','Arial','fontsize', 12);
ylabel('MSC', 'FontName','Arial','fontsize', 12);
subplot(1,2,2)
plot(MMSC_Index,'rd','LineStyle','-','LineWidth',1.3,
'MarkerSize',5);
xlabel('Node No.', 'FontName','Arial','fontsize', 12);
ylabel('MMSC', 'FontName','Arial','fontsize', 12);
xlim([1 32])
set(gcf,'Position',[200 200 1000 240]);
```

4.4 Summary

This chapter demonstrated the application of the computer vision sensor for experimental modal analysis, structural model updating, and damage detection/location, through example laboratory experiments. The experimental modal analysis of a three-story frame structure and a beam structure showed that the identified natural frequencies and mode shapes from measurements using one camera tracking multiple existing natural targets on the structure agree well with those obtained using multiple accelerometers and/or laser displacements, with the additional benefit of increased spatial resolution in the identified mode shapes. An approach for structural model updating based on experimental modal analysis (using computer vision multipoint measurement) and frequency-domain optimization was described and tested on the three-story frame structure, successfully updating its inter-story stiffness values. In addition, a damage-detection method based on MSC was presented and applied to the beam structure. The unique advantage of the dense multipoint measurement of the computer vision sensor enabled the construction of a high-resolution MSC index, resulting in successful damage detection with accurate damage location. Compared with conventional pointwise on-structure sensors, the noncontact computer vision sensor can achieve high spatial resolution multipoint displacement measurements and has proven to be cost-effective for experimental modal analysis, structural parameter identification and modal updating, and structural damage detection and location.

5

Application in Model Updating of Railway Bridges under Trainloads

The network in the United States is dominated by freight railroads, which transport over 40% of the nation's intercity freight. The freight railroad network, including 76 000 bridges, was largely built 100 years ago, and most of these bridges are still in service. Over the last few decades, traffic loads have increased due to higher transport efficiency and demand from the continuously growing economy, subjecting these bridges to loads much greater than what they were designed to carry. This accelerates the deterioration process of the aging bridge structures and poses inspection, maintenance, and management challenges. It is of particular interest to closely monitor the displacement of railroad bridges under trainloads, as excessive displacement not only accelerates fatigue in bridge structures but can potentially cause track instability and loss of contact between the rail and train wheels. It is highly challenging, if not impossible, to install contact-type sensors such as a linear variable differential transformer (LVDT) or a string potentiometer on a bridge crossing a river with high piers, due to the difficulty of connecting the sensor to a stationary reference point. The noncontact computer vision sensor has demonstrated its significant advantages in cost-effective measurement of actual railway bridges, as presented in Section 3.7.

This chapter further discusses how the measured displacement responses can be used to update analytical models of railway bridges. As discussed in Chapter 4, the existing model-updating methods in the literature are often based on experimental modal analysis using the measured input excitation and output structural responses. To update railway bridge models, Ribeiro et al. [46] conducted an ambient vibrational test and applied a genetic algorithm to update 15 parameters of the model. In addition, Wiberg et al. [47] updated the model of a railway bridge based on operational modal analysis and loading tests.

Computer Vision for Structural Dynamics and Health Monitoring, First Edition.
Dongming Feng and Maria Q. Feng.
© 2021 John Wiley & Sons Ltd.
This Work is a co-publication between John Wiley & Sons Ltd and ASME Press.
Companion website: www.wiley.com/go/feng/structuralhealthmonitoring

However, model-updating methods based on ambient vibration measurement and modal analysis suffer from technical challenges, particularly for short-span railway bridges (the majority in the US freight railway network), which are extremely stiff and experience little ambient vibrations, making it difficult to accurately extract modal parameters. Therefore, it is desirable to update the model using in situ vibration measurements under trainloads for short-span railway bridges. But a railway bridge coupled with a moving train is a time-varying system with changing frequencies during the train's passage [48]. In the measured railway bridge response, the natural frequencies of the bridge are overshadowed and hidden due to their much smaller amplitudes compared with those of the dominant frequencies induced by repeated trainloads (discussed later in this chapter). It would be incorrect to assume white-noise excitation and use the bridge output only to identify the bridge's modal properties and update its analytical model via the frequency-domain optimization, as presented in Chapter 4.

This chapter presents a time-domain finite element approach to updating models for railway bridges based on the train-induced bridge displacement response as measured by the computer vision sensor and given knowledge about the trainloads. Section 5.1 describes the in situ vision-based displacement measurement of a short-span plate girder bridge, a type of bridge frequently used in the railway bridge network. Section 5.2 develops an initial finite element model that takes into consideration the train-track-bridge interaction. In Section 5.3, parameter sensitivity analysis is performed to demonstrate that displacement measurement is more suited than the commonly used acceleration measurement to update bridge stiffness, making this scenario well-suited for the computer vision sensor, which directly measures displacement. Then a two-step model-updating procedure is presented and applied to the bridge to update its stiffness together with the train speed. In Section 5.4, the dynamic effects induced on the short-span railway bridge by heavy freight trains are investigated from the analysis of the frequency characteristics of the train-bridge interaction system.

5.1 Field Measurement of Bridge Displacement under Trainloads

A 100-year-old testbed steel bridge, located at Pueblo, Colorado (as described in Chapter 3), is considered for model updating. As shown in Figure 5.1, this short-span simply supported bridge is 16.75 m long, consisting of two I-shape riveted plate girders and the horizontal and vertical brace system, sleepers, and rail system. Field measurements are carried out on the bridge under a moving train, which has one locomotive weighting 89 586 kg and 15 freight cars weighing 71 440 kg each. Each of the freight cars is approximately 12 times heavier than the bridge's weight. The dimensions of the train are indicated in Figure 5.2.

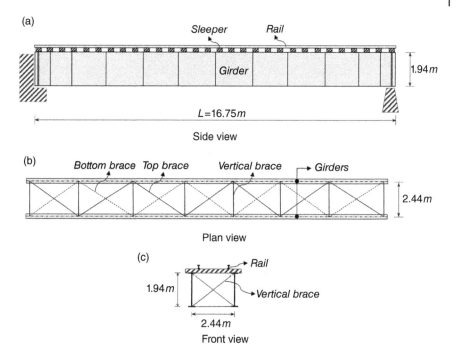

Figure 5.1 Railway bridge for model updating: (a) side view; (b) plan view; (c) front view. *Source:* Reproduced with permission of ASCE Library.

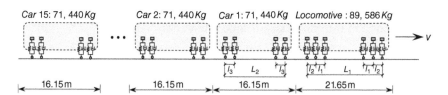

Figure 5.2 Freight train configuration. *Source:* Reproduced with permission of ASCE Library.

The field tests focus on measuring the mid-span vertical displacement of the bridge using the vision-based sensor, as discussed in Chapter 3. The tests are carried out with various train speeds, i.e. approximately 8.05 km/h (5 mph), 38.62 km/h (24 mph), and 64.36 km/h (40 mph), as measured by the train speedometer. The vision sensor system is placed at a stationary position 9.15 m from the bridge. A typical displacement time history measured at the mid-span under 8.05 km/h trainloads is shown in Figure 5.3. A negative sign represents a downward

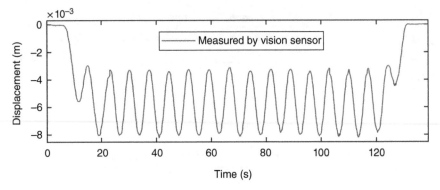

Figure 5.3 Displacement history with train speed 8.05 km/h.

deflection. The displacement history reflects the oscillating trainload pattern, with 17 peaks matching the number of cars. The contribution of each car is clearly distinguished.

5.2 Formulation of the Finite Element Model

An initial finite element (FE) model is formulated for this railway bridge, taking into consideration the interactions among the three subsystems: the train, track, and bridge. This model serves as a baseline for updating in Section 5.3.

5.2.1 Modeling the Train-Track-Bridge Interaction

Considerable efforts have been made toward the development of efficient numerical vehicle-bridge interaction models to predict dynamic responses of railway and highway bridges. For railway bridges, these studies may be generally divided into three categories based on techniques used to model trainloads: the moving load model, the moving mass model, and the moving spring-damper system model. Of the three models, the moving load model is the simplest and the most computationally efficient, but the train-bridge interaction effect is ignored. The moving mass model takes into account the train inertia effect but does not consider the bouncing action of the moving train on the bridge, which is expected to be significant for trains moving at high speeds or in the presence of track roughness. In view of this, Cheng et al. [49] proposed a bridge-track-vehicle element to study the dynamic responses of railway bridges. The moving train is modeled as a series mass-spring-damper systems at the axle locations with two degrees of freedom (DOFs), and thus vibration of the tracks can also be simultaneously analyzed.

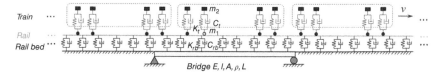

Figure 5.4 Schematic representation of the bridge-track-vehicle interaction system.

In this study, the dynamic response of the railway bridge in the vertical direction is considered without rolling and yawing effects. Thus the train-track-bridge interaction system is idealized by a 2D FE model, as depicted in Figure 5.4. In the model, the upper and lower beam elements are used to model the rail and the bridge girder, respectively. The elasticity and damping properties of the rail bed are represented by a series of springs and viscous dampers. The track on the embankment is modeled as a beam on a viscoelastic foundation.

The train subsystem model adopts the following assumptions: (i) the train runs on the bridge at a constant speed; (ii) the train can be modeled as several (in this case, six for the locomotive and four for each of the cars) independent spring-damper suspension systems at the axle locations; (iii) only the vertical DOF is considered; thus each spring-damper suspension system has two DOFs.

By combining the equations of motion for each of the three subsystems, the equation of motion of the train-track-bridge system can be derived in sub-matrix form as

$$
\begin{bmatrix} \mathbf{M}_{bb} & \mathbf{0} & \mathbf{0} \\ \mathbf{0} & \mathbf{M}_{rr} & \mathbf{0} \\ \mathbf{0} & \mathbf{0} & \mathbf{M}_{tt} \end{bmatrix} \begin{Bmatrix} \ddot{\mathbf{X}}_b \\ \ddot{\mathbf{X}}_r \\ \ddot{\mathbf{X}}_t \end{Bmatrix} + \begin{bmatrix} \mathbf{C}_{bb} & \mathbf{C}_{br} & \mathbf{0} \\ \mathbf{C}_{rb} & \mathbf{C}_{rr} & \mathbf{C}_{rt} \\ \mathbf{0} & \mathbf{C}_{tr} & \mathbf{C}_{tt} \end{bmatrix} \begin{Bmatrix} \dot{\mathbf{X}}_b \\ \dot{\mathbf{X}}_r \\ \dot{\mathbf{X}}_t \end{Bmatrix}
$$
$$
+ \begin{bmatrix} \mathbf{K}_{bb} & \mathbf{K}_{br} & \mathbf{0} \\ \mathbf{K}_{rb} & \mathbf{K}_{rr} & \mathbf{K}_{rt} \\ \mathbf{0} & \mathbf{K}_{tr} & \mathbf{K}_{tt} \end{bmatrix} \begin{Bmatrix} \mathbf{X}_b \\ \mathbf{X}_r \\ \mathbf{X}_t \end{Bmatrix} = \begin{Bmatrix} \mathbf{F}_b \\ \mathbf{F}_r \\ \mathbf{F}_t \end{Bmatrix} \tag{5.1}
$$

where the subscripts b, r, and t denote, respectively, the bridge, the rail track, and the train. The boldfaced vectors $\mathbf{X}_b$, $\mathbf{X}_r$, and $\mathbf{X}_t$ denote the displacement vectors of the bridge, the rail, and the train axles, respectively. The matrices $\mathbf{M}_{tt}$, $\mathbf{K}_{tt}$, and $\mathbf{C}_{tt}$ of the train are marked with the subscript tt. The matrices $\mathbf{M}_{rr}$, $\mathbf{K}_{rr}$, and $\mathbf{C}_{rr}$ of the rail are marked with the subscript rr. The matrices $\mathbf{M}_{bb}$, $\mathbf{K}_{bb}$, and $\mathbf{C}_{bb}$ of the bridge are marked with the subscript bb. The matrices $\mathbf{K}_{rt}$, $\mathbf{C}_{rt}$, $\mathbf{K}_{tr}$, and $\mathbf{C}_{tr}$ induced by the train-rail interaction are denoted by the subscript rt or tr. The matrices $\mathbf{K}_{br}$, $\mathbf{C}_{br}$, $\mathbf{K}_{rb}$, and $\mathbf{C}_{rb}$ induced by the bridge-rail interaction are denoted by the subscript br or rb.

This equation can be further written as

$$\mathbf{M}(t)\ddot{\mathbf{X}}(t) + \mathbf{C}(t)\dot{\mathbf{X}}(t) + \mathbf{K}(t)\mathbf{X}(t) = \mathbf{F}(t) \tag{5.2}$$

where $\mathbf{M}(t)$, $\mathbf{C}(t)$, and $\mathbf{K}(t)$ are the global mass, damping, and stiffness matrices, respectively; $\mathbf{X}(t)$, $\dot{\mathbf{X}}(t)$, and $\ddot{\mathbf{X}}(t)$ are the displacement, velocity, and acceleration vectors, respectively; and $\mathbf{F}(t)$ is the force vector. Note that the interaction system represents a coupled time-varying dynamic system: the mass, damping, and stiffness matrices as well as the force vectors of the system change with time due to interaction with the moving train. Therefore they must be updated at each time step. By using the Newmark-β method, Eq. (5.1) or Eq. (5.2) can be solved step by step to obtain the dynamic responses of the bridge, the train, and the track simultaneously.

Here, as a special case of Rayleigh damping, stiffness-proportional damping is assumed for the bridge, defined as

$$\mathbf{c}_b = \alpha \mathbf{k}_b \tag{5.3}$$

where $\mathbf{c}_b$ and $\mathbf{k}_b$ are the damping and stiffness matrices of the bridge beam element, respectively; and α is the constant proportional coefficient and set to 0.01.

5.2.2 Finite Element Model of the Railway Bridge

Tables 5.1 and 5.2 list the parameter values of the bridge, rail system, and freight train subsystem for the initial FE model. The values for the bridge structural parameters are based on the design drawings. The elasticity and damping properties of the rail bed and the train suspension systems, i.e. K_{rb}, C_{rb}, K_t, and C_t, are unavailable; values are approximated according to the literature [50]. For a 2D model studying the vertical dynamic response, only half of the train-track-bridge system is considered due to its symmetry. There are a total of 3138 DOFs for the train-track-bridge FE model. Figure 5.5 compares the measured displacement with the simulated displacement using the initial baseline FE model under

Table 5.1 Design parameters of the bridge and track system.

Item	E (N/m^2)	I (m^4)	A (m^2)	ρ (Kg/m^3)	υ	Stiffness (N/m)	Damping (Ns/m)
Girder	2.10e11	3.69e-2	5.15e-2	7850	0.30	—	—
Rail	2.10e11	6.12e-5	1.54e-2	7850	0.30	—	—
K_{rb}	—	—	—	—	—	5.00e8	—
C_{rb}	—	—	—	—	—	—	2.00e5

Table 5.2 Parameters of the freight train.

Item	Unit	1 locomotive	15 freight cars
Axle space	m	$l_1 = 2.13; l_2 = 1.41; L_l = 16.80$	$l_3 = 1.83; L_2 = 14.17$
Wheel-set m_1	Kg	500	500
Body m_2	Kg	14431	17360
Stiffness K_t	N/m	1.45e6	1.45e6
Damping C_t	N/m	3.00e4	3.00e4

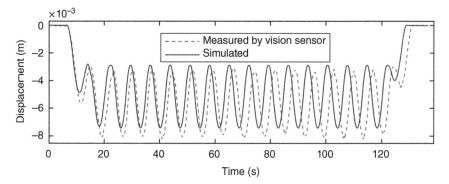

Figure 5.5 Measured vs. simulated displacement using the initial FE model.

trainloads with a speed of 8.05 km/h. Some disagreements in amplitude and time delay are observed. This is considered to be caused by uncertainties in the adopted system parameters and the train speed and demonstrates the need to update the model using in situ measurement data.

5.3 Sensitivity Analysis and Finite Element Model Updating

Traditionally, acceleration is utilized for dynamic tests of bridges, because it is more difficult to install conventional displacement sensors such as an LVDT due to the need for a stationary reference point. Since the emerging computer vision sensor enables cost-effective measurement of structural dynamic displacement, it becomes of interest to compare the modal-updating performance using displacement and acceleration response measurements. A sensitivity analysis is carried out to compare the influence of the various model parameters on the bridge displacement and acceleration responses. Based on the results, a two-step approach for updating models is presented and applied to the railway bridge.

5.3.1 Model Updating as a Time-Domain Optimization Problem

The FE model-updating process is essentially an optimization problem in which the model parameters with the highest uncertainty are updated to reflect the current structural condition based on measured responses. While Section 4.3 presents an objective function that minimizes the discrepancies between computed and measured modal properties, including natural frequencies and mode shapes, the objective function here is defined as the distance between the computed and measured displacement time history data. Consider that $\mathbf{u}(t_k)$ and $\mathbf{y}(t_k)$ are the true input and output vectors, respectively, of a structural system at time instant t_k, where $k = 1, 2, \cdots, n_{time}$, and that the physical system can be described by a set of parameters assembled in a parameter vector $\boldsymbol{\theta} = \{\theta_1, \theta_2, \cdots, \theta_n\}^T$. In a structural system, the unknown parameters θ_i $(i = 1, 2, \cdots, n)$ could be the mass, damping, or stiffness. Then the input–output relationship can be represented by this general expression:

$$\mathbf{y}(t_k) = f\left(\mathbf{u}(t_k), \boldsymbol{\theta}\right) \qquad (5.4)$$

Denote $\mathbf{Y}(t_k)$ as the estimated structural response obtained from the FE model using the same input $\mathbf{u}(t_k)$ and an estimated set of the structural parameter values $\boldsymbol{\Theta} = \{\Theta_1, \Theta_2, \cdots, \Theta_n\}^T$:

$$\mathbf{Y}(t_k) = f\left(\mathbf{u}(t_k), \boldsymbol{\Theta}\right) \qquad (5.5)$$

The objective of the model-updating procedure is to find the best estimates of $\boldsymbol{\Theta}$ so as to minimize the discrepancy between the measured response $\mathbf{y}(t_k)$ and the predicted $\mathbf{Y}(t_k)$ over the entire time history. In problems that involve measuring dynamic responses, the objective function to be minimized through the optimization process can be formulated as the sum of the mean square error (MSE) between the measured and predicted responses over the measurement period, such as

$$\Pi(\boldsymbol{\Theta}) = \frac{1}{n_{rec} n_{time}} \sum_{j=1}^{n_{rec}} \sum_{k=1}^{n_{time}} \left\| Y_j(t_k) - y_j(t_k) \right\| \qquad (5.6)$$

where n_{rec} is the number of recorded time histories, which depends on the number of sensors used in the tests and; $\|\cdot\|$ denotes the Euclidean norm of a vector. The overall model-updating problem is then summarized as follows:

$$\text{Find } \boldsymbol{\Theta} = \{\Theta_1, \Theta_2, \cdots, \Theta_n\}^T \in \Gamma \text{ such that } \Pi(\boldsymbol{\Theta}) \text{ is minimized} \qquad (5.7)$$

where Γ is the feasible n-dimensional parameter search space

$$\Gamma = \left\{ \Gamma \in \mathbb{R}^n \middle| \theta_i^{\min} \le \Theta_i \le \theta_i^{\max}, i = 1, 2, \cdots, n \right\} \tag{5.8}$$

where n is the number of parameters to be identified and $\theta_i^{\min}$ and $\theta_i^{\max}$ are the lower and upper bounds of the ith parameter. In summary, the problem of updating an FE model is treated as a linearly constrained nonlinear optimization problem in this study.

5.3.2 Sensitivity Analysis of Displacement and Acceleration Responses

The dynamic response of the bridge is influenced by the various parameters of the bridge, train, and track subsystems in the analytical model in Eq. (5.2). In an attempt to gain insight into the contribution of each of the parameters to the objective function, a sensitivity analysis is carried out focusing on the following six nondimensional normalized parameters: bridge stiffness R_{EI}, bridge damping R_α, rail bed stiffness $R_{K_{rb}}$, rail bed damping $R_{C_{rb}}$, train stiffness R_{K_t}, and train damping R_{C_t}:

$$
\begin{aligned}
R_{EI} &= E'I' / EI, & 0.5 \le R_{EI} \le 0.5 \\
R_\alpha &= \alpha' / \alpha, & 0.5 \le R_\alpha \le 0.5 \\
R_{K_{rb}} &= K'_{rb} / K_{rb}, & 0.1 \le R_{K_{rb}} \le 10 \\
R_{C_{rb}} &= C'_{rb} / C_{rb}, & 0.1 \le R_{C_{rb}} \le 10 \\
R_{K_t} &= K'_t / K_t, & 0.1 \le R_{K_t} \le 10 \\
R_{C_t} &= C'_t / C_t, & 0.1 \le R_{C_t} \le 10
\end{aligned} \tag{5.9}
$$

are functions of the parameters in the initial model as discussed in Section 5.3.2 (equivalent bridge stiffness EI, stiffness-proportional Rayleigh damping coefficient α [recall that α is the constant factor in Eq. (5.3)], rail bed stiffness K_{rb} and damping C_{rb}, train suspension stiffness K_t and damping C_t) and corresponding parameter variations in a specified range ($E'I'$, α', K'_{rb}, C'_{rb}, K'_t and C'_t). As shown in Eq. (5.9), $E'I'$ and α' vary in a range from 0.5 to 1.5 – a maximum of a 50% decrease or increase relative to the reference values of EI and α. In reality, the values of K'_{rb}, C'_{rb}, K'_t, and $C'_t$are prone to relatively higher uncertainties and are given a larger variation range from 0.1 to 10 (a maximum decrease or increase of 10 times their corresponding reference values).

The procedure for the sensitivity analysis is illustrated in Figure 5.6. The bridge displacement and acceleration time histories at the mid-span point are computed using the formulated model with the parameter variations in Eq. (5.9), respectively, with train speeds $v = 16.09, 32.18, 48.27, 64.36, 80.45, 96.54$, and $112.63\,\text{km/h}$

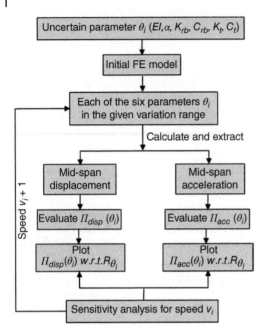

Figure 5.6 Sensitivity analysis procedure. *Source:* Reproduced with permission of ASCE Library.

(or 10, 20, 30, 40, 50, 60, 70 mph). The speed is chosen based on the maximum possible running speed limit of freight trains of 129 km/h (80 mph) on a class 5 track, according to the Federal Railroad Administration. Then, the objective function defined in Eq. (5.6) is used to evaluate curves $\Pi_{disp}(\theta_i)$ and curves $\Pi_{acc}(\theta_i)$, in which $\mathbf{y}(t_k)$ represents the displacement or acceleration response obtained using the baseline parameter values and $\mathbf{Y}(t_k)$ using the varied parameter values. The larger the objective function value of a specified parameter, the more sensitive the modeled output is to that parameter. Figures 5.7–5.12 plot the values of the displacement and acceleration objective functions with respect to (w.r.t.) each of the six parameters at the given variation ranges for different freight train speeds. Note that the x-axes in Figures 5.9–5.12 are in log scale. In total, 742 sets of FE dynamic response simulations are carried out for this sensitivity analysis.

The following observations can be made from the sensitivity analysis results, as shown in the figures:

- *Displacement sensitivity* – The sensitivity of the bridge displacement response to the change in bridge equivalent stiffness EI as represented by $\Pi_{disp}(\theta_i)$ is several orders of magnitude greater than the other five parameters (α, K_{rb}, C_{rb}, K_t, and C_t, mostly associated with the train and rail bed) for the same parameter variation range, such as -0.5 to $+0.5$. The displacement sensitivity to bridge stiffness

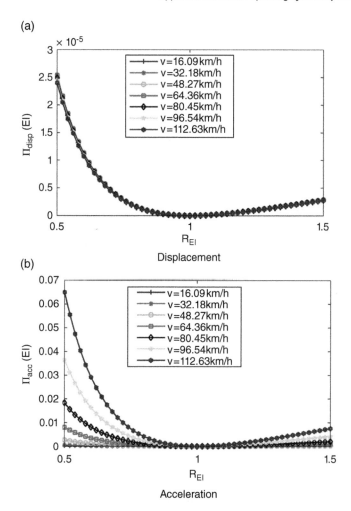

Figure 5.7 Objective functions w.r.t. normalized bridge equivalent stiffness R_{EI}: (a) displacement; (b) acceleration. *Source:* Reproduced with permission of ASCE Library.

is hardly affected by train speed, while the acceleration sensitivity to bridge stiffness depends on the train speed. The fact that the bridge equivalent stiffness dominates the parameter contributions to the bridge displacement response makes the displacement measurement the best metric to update the bridge stiffness. Additionally, assuming the plate bridge has a constant EI along its girder, only the vertical displacement at one point, such as the mid-span point, needs to be measured in order to update the EI.

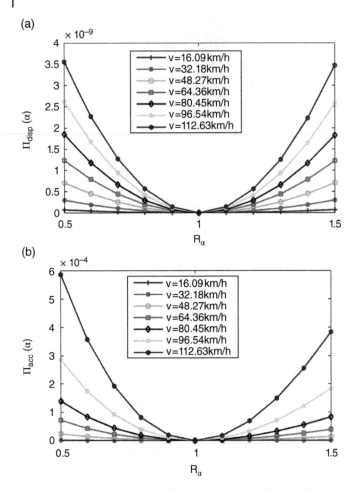

Figure 5.8 Objective functions w.r.t. normalized bridge damping R_α: (a) displacement; (b) acceleration. *Source:* Reproduced with permission of ASCE Library.

- *Acceleration sensitivity* – The bridge acceleration response to trainloads is affected by several parameters. Unlike displacement, the sensitivity of the bridge acceleration response to bridge stiffness, as represented by $\Pi_{acc}(\theta_i)$, is not significantly greater than the other parameters. Moreover, the acceleration sensitivity to bridge stiffness is highly dependent on the train speed; and at a low train speed, acceleration is not sensitive to bridge stiffness. Therefore, to identify or update the bridge girder stiffness, the bridge displacement response measurement is more effective than the acceleration measurement.

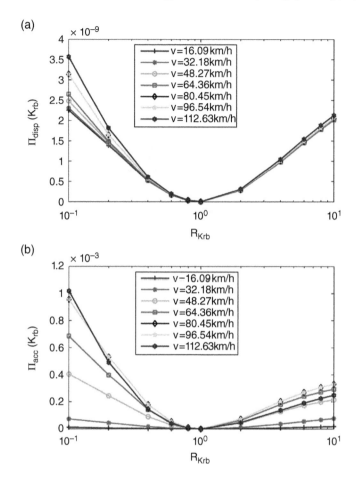

Figure 5.9 Objective functions w.r.t. normalized rail bed stiffness $R_{K_{rb}}$: (a) displacement; (b) acceleration. *Source:* Reproduced with permission of ASCE Library.

5.3.3 Finite Element Model Updating

Based on the sensitivity analysis, an approach for updating FE models is formulated for short-span railway bridges based on measuring the bridge displacement response to trainloads. As shown in Figure 5.13, the approach involves two steps to identify the train speed and the equivalent bridge stiffness *EI*. The identification is based on time-domain optimization, as presented in Section 4.1. The iterative Nelder–Mead simplex method, an optimization algorithm incorporated in MATLAB, is adopted to minimize the objective function in Eq. (5.6). This algorithm does not require the use of the gradient or Hessian of the objective function and is less prone to numerical difficulties during iteration.

(a)

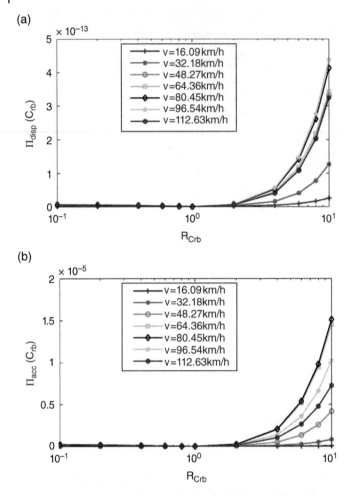

(b)

Figure 5.10 Objective functions w.r.t. normalized rail bed damping $R_{C_{rb}}$ (a) displacement; (b) acceleration. *Source:* Reproduced with permission of ASCE Library.

1) *Update the train speed.* Only the train speed v is updated, and v is set as the unknown. Equation (5.6) is rewritten as $\Pi(v)$. The initial value of the train speed (8.05 km/h, read from the train speedometer) is updated to be 7.93 km/h. Figure 5.14 compares the measured bridge mid-span vertical displacement with that computed using the updated train speed. Compared with the plot in Figure 5.5 before updating, the time delay between the measured and computed displacement time histories is reduced. The amplitude disagreement, caused primarily by the uncertain bridge stiffness EI, is corrected in Step 2.

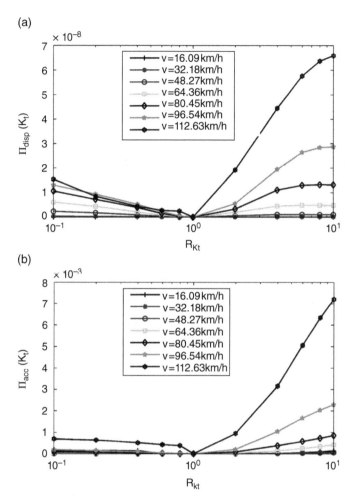

Figure 5.11 Objective functions w.r.t. normalized train suspension stiffness R_{K_t}: (a) displacement; (b) acceleration. *Source:* Reproduced with permission of ASCE Library.

2) *Update the equivalent bridge stiffness (EI).* Plugging in the updated train speed $v^{updated}$, the objective function in Eq. (5.6) becomes $\Pi(v^{updated}, EI)$, which is used to identify the bridge stiffness. The initial stiffness value $EI = 7.75 \times 10^9 Nm^2$, based on the design document (as listed in Table 5.1), is updated to $EI = 7.09 \times 10^9 Nm^2$. After the update, the computed displacement time history better matches the measured value, as shown in Figure 5.15 (in comparison with the baseline in Figure 5.5).

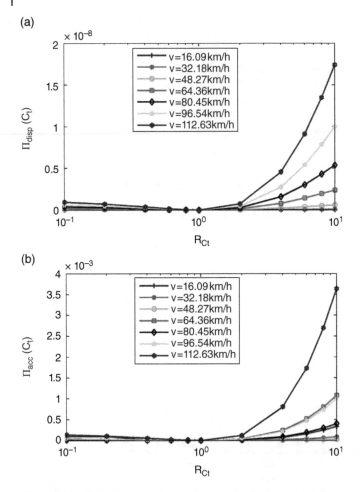

Figure 5.12 Objective functions w.r.t. normalized train suspension damping R_{C_t}: (a) displacement; (b) acceleration. *Source:* Reproduced with permission of ASCE Library.

5.4 Dynamic Characteristics of Short-Span Bridges under Trainloads

Railway bridges exhibit unique frequency characteristics under trainloads, due to the interaction of the bridge and the heavy moving trains. It is of research interest and is also a practical need to investigate whether the updated analytical model can capture the dynamic characteristics of the coupled train-track-bridge system.

As introduced in the literature [48, 51], for a typical train passage with multiple wheel sets on each car, the bridge's response is dominated by trainload excitation

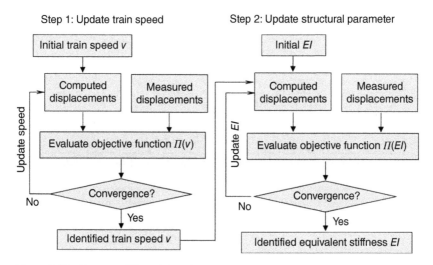

Figure 5.13 Two-step FE model-updating procedure. *Source:* Reproduced with permission of ASCE Library.

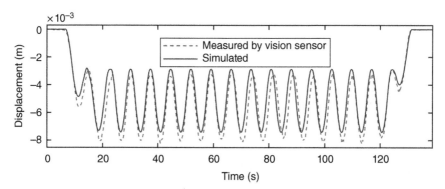

Figure 5.14 After Step 1: train speed update. *Source:* Reproduced with permission of ASCE Library.

frequencies. The equally spaced dominant frequencies arising from repeated trainloads can be expressed as

$$f_{dominant} = nv / l_{train}, \quad n = 1,2,3\cdots \tag{5.10}$$

where n is the order of the dominant frequency; v is the train speed; and l_{train} is the distance between two car centers, as shown in Figure 5.16. That is, the dominant frequency $f_{dominant}$ depends only on the train's speed and car length.

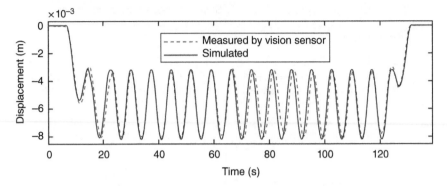

Figure 5.15 After Step 2: equivalent bridge stiffness update. *Source:* Reproduced with permission of ASCE Library.

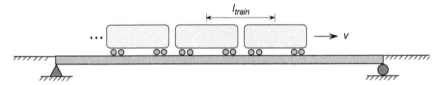

Figure 5.16 Bridge under a moving train.

According to Eq. (5.10), the first-order excitation frequency $f_{dominant}$ of the train-loads is expected to be 0.136, 0.632, and 1.207 Hz given the three tested train speeds, 7.93, 36.80, and 70.22 km/h (or the updated speeds, 8.05, 38.62, and 64.36 km/h).

On the other hand, Figure 5.17 plots the power spectral density (PSD) computed from the measured mid-span displacements at the three train speeds. The peak frequencies in the PSDs given the three train speeds are 0.137, 0.631, and 1.2 Hz, respectively, which perfectly agree with the first-order frequencies of the trainloads with the same train speeds. This demonstrates that the bridge response displacement is dominated by the first-order frequency of the trainloads, which depends on the train speed.

Furthermore, Figures 5.18–5.20 show the displacement and acceleration time histories, together with their corresponding PSDs, computed using the updated FE model. Again, only the first-order excitation frequency of the trainloads is clearly identified in the bridge response for each of the three different speeds, while the higher-order excitation frequencies and the natural frequencies of the bridge are overshadowed and hidden due to their much smaller amplitudes compared with those of the trainloads' first dominant frequency.

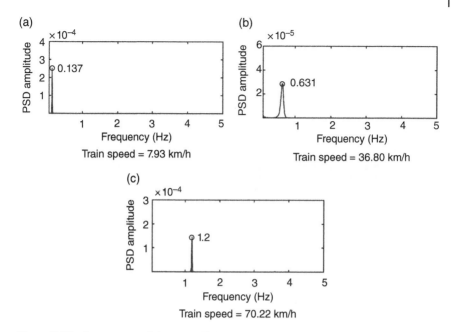

Figure 5.17 Power spectral density (PSD) of measured displacement histories: (a) train speed = 7.93 km/h; (b) train speed = 36.80 km/h; (c) train speed = 70.22 km/h. *Source:* Reproduced with permission of ASCE Library.

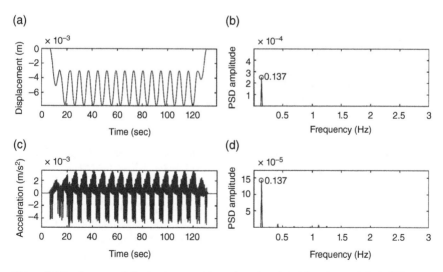

Figure 5.18 Computed displacement and acceleration time histories and their PSDs with train speed 7.93 km/h. *Source:* Reproduced with permission of ASCE Library.

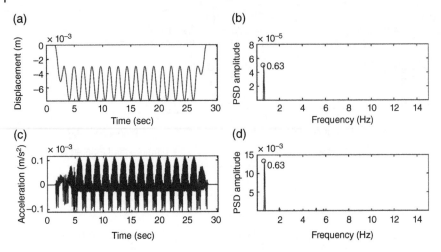

Figure 5.19 Computed displacement and acceleration time histories and their PSDs with train speed 36.80 km/h. *Source:* Reproduced with permission of ASCE Library.

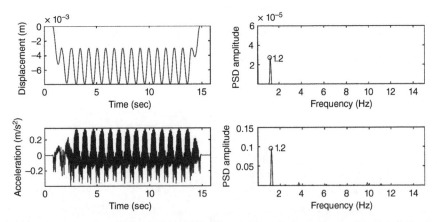

Figure 5.20 Computed displacement and acceleration time histories and their PSD with train speed 70.22 km/h. *Source:* Reproduced with permission of ASCE Library.

Table 5.3 compares the dominant frequency values given the three train speeds obtained from (i) the input excitation computed by Eq. (5.10) using the updated train speed, (ii) the measured output bridge response displacement, and (iii) the output bridge displacement computed using the updated FE model (i.e. the updated stiffness value). The table confirms that (i) the response displacement of the bridge is dominated by the first-order excitation frequency of the trainload and (ii) the dominant frequency increases as the train speed increases.

Table 5.3 Dominant frequencies.

$f_{dominant}$ (Hz)	Train speed (km/h)		
	$v = 7.93$	$v = 36.80$	$v = 70.22$
Excitation	0.136	0.632	1.207
Response (measured)	0.137	0.631	1.200
Response (computed)	0.137	0.630	1.200

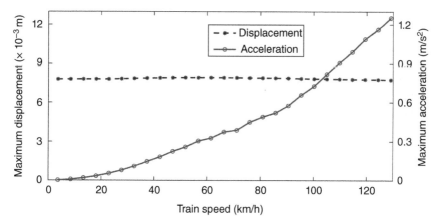

Figure 5.21 Mid-span maximum displacements and accelerations w.r.t. different train speeds. *Source:* Reproduced with permission of ASCE Library.

While the bridge displacement response frequency depends on the train speed (as the displacement is dominated by the trainloads' excitation frequency), the bridge response displacement amplitudes remain constant regardless of the train speed. On the other hand, the bridge acceleration response amplitudes increase as the train speed increases due to the train-track-bridge dynamic interactions. To gain more insight into the dynamic effects of the short-span railway bridge, the updated FE model is utilized to compute dynamic responses for train speeds in a wide range: 3.60–129 km/h (2.24–80 mph), at increments of 3 km/h. Figure 5.21 shows the maximum values of the mid-span displacement and acceleration time histories. As noted earlier, the maximum displacement remains almost constant (about 7.8 mm) regardless of the train speed, while maximum acceleration increases significantly as the train speed increases, as a result of train-track-bridge interaction.

The first natural frequency of the short-span bridge is calculated as 24.53 Hz from the FE model, which is considerably higher than the excitation frequencies of the repeated trainloads with energy concentrated on the first dominant excitation frequency. As a result, the displacement response of the bridge is dominated by this first-order excitation frequency of the trainloads.

This example demonstrates that the natural frequencies of the railway bridges cannot be identified from only train-induced dynamic responses without knowledge regarding the trainloads. For common stiff, short-span railway bridges with natural frequencies much higher than the dominant frequency of the heavy freight trainloads, the bridges' natural frequencies are not excited by trainloads; in addition, the bridge displacement decays to zero immediately after the train leaves the bridge, as shown in Figure 5.3, making it impossible to identify the bridge's natural frequencies from free vibrations. For such stiff bridges, it is also difficult to conduct experimental modal analysis using low-amplitude ambient vibrations. For this reason, the time-domain-based optimization approach presented in this chapter can be effectively applied to update models in lieu of the frequency-domain approach discussed in Chapter 4. As discussed in the sensitivity study in Section 5.3.2, the displacement response amplitude is significantly more sensitive to the bridge's equivalent stiffness than other parameters in the train-track-bridge system. Therefore, the measured displacement time history data is particularly useful for identifying and updating the bridge stiffness value in the FE model.

5.5 Summary

To address the need for monitoring aging railway bridges, this chapter presented a time-domain approach to update models using train-induced bridge displacement responses measured by the computer vision sensor. The model-updating procedure was carried out in two steps to update the train speed and the bridge's equivalent stiffness. The approach was validated on a short-span railway bridge by measuring bridge displacements under moving trainloads using the computer vision sensor.

From the frequency domain analysis of both the in situ measurement and the simulation analysis, it was found that the dominant frequency of the bridge displacement response was consistent with the trainloads' first-order excitation frequency. In other words, the bridge response was dominated by the excitation frequency associated with the train passing. For stiff, short-span railway bridges, the bridge's natural frequencies are much higher than the trainload's excitation frequency and thus are not excited by trainloads. It would be incorrect to model trainloads as white noise excitations and identify the bridge's natural frequencies from bridge responses only. The proposed model-updating method based on time-domain optimization addresses this difficulty. To implement this method, knowledge of the trainloads is required. Note that the train speed can be directly estimated from video images captured by the computer vision system, although it was identified through the optimization procedure in this study.

The sensitivity analysis revealed that the bridge's dynamic displacement response amplitude was significantly more sensitive to the bridge equivalent stiffness EI than other parameters such as bridge damping, rail bed stiffness and damping, and train suspension stiffness and damping. This makes train-induced bridge displacement measurement (which can be conveniently made by the computer vision sensor) particularly useful for identifying and updating bridge girder stiffness in the finite element model. Such an updated analytical model can serve as a baseline for monitoring long-term structural health and detecting damage.

6

Application in Simultaneously Identifying Structural Parameters and Excitation Forces

Identifying structural parameters can be formulated as a frequency- or time-domain optimization process, aiming to minimize the discrepancy between the measured and computed structural response time histories as described in Chapters 4 and 5. Measured or prior knowledge about the input forces is required for both approaches. For example, the identification of the railway bridge stiffness in Chapter 5 is based on prior knowledge about the trainloads such as the car length and mass and the train speed. For a majority of structures, however, the input excitation is unknown and highly difficult, if not impossible, to measure.

Therefore, it is of interest to develop techniques for simultaneously identifying both structural parameters and input force time histories. Accurately identifying the unknown forces applied on the structure is also beneficial for proactive structural maintenance and management. However, most existing methods for force identification in the time domain are based on the assumption that the system characteristics are "completely known": i.e. that there is a full-scale finite element model or reduced-order model available, together with its parameter values. Many researchers have worked on this inverse problem of identifying structural input forces. Hollandsworth and Busby [52] studied the impact force for a cantilever beam structure and found that an accelerometer located closer to the location of the impact force could provide better identification of the force-time history. Inoue et al. [53] adopted the least squares method based on singular value decomposition to estimate the magnitude and direction of impact force, which was demonstrated on a simply supported beam using strain responses measured by strain gauges. Wang and Chiu [54] developed both time and frequency domain prediction methods for determining the location and amplitude of an unknown impact force acting on a simply supported beam by minimizing the objective function defined by the sum of square errors between the predicted and measured

Computer Vision for Structural Dynamics and Health Monitoring, First Edition.
Dongming Feng and Maria Q. Feng.
© 2021 John Wiley & Sons Ltd.
This Work is a co-publication between John Wiley & Sons Ltd and ASME Press.
Companion website: www.wiley.com/go/feng/structuralhealthmonitoring

acceleration responses, with the amplitude and the location of the unknown impact force defined as design variables. Inoue et al. [55] presented a review of inverse analysis methods for the indirect measurement of impact forces.

This chapter presents a new method for simultaneously identifying the input force and structural parameters by utilizing structural displacement response time histories measured by the computer vision sensor. Section 6.1 introduces this output-only identification method based on an iterative parametric optimization process. In Section 6.2, a numerical example is presented to evaluate the accuracy of the method, in which the effects of displacement measurement noise, number of sensors, and initial values of structural parameters on the identification accuracy are investigated. Section 6.3 presents an experimental validation of this simultaneous identification method.

6.1 Simultaneous Identification Using Vision-Based Displacement Measurements

This section introduces the formulation of the simultaneous identification of unknown structural parameters and input forces using output displacement time histories measured by the vision sensor. The concept is illustrated in Figure 6.1.

Note that while the examples presented in this chapter are beam structures subjected to an impact force, the simultaneous identification method is applicable to other types of structures and excitation forces beyond the impact force (such as seismic ground motion). The computer vision sensor makes it easier to implement this method.

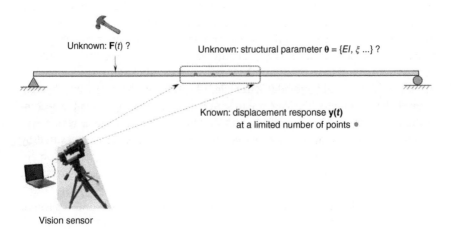

Figure 6.1 Schematics of the output-only simultaneous identification problem.

6.1.1   Structural Parameter Identification as a Time-Domain Optimization Problem

As discussed in Chapter 5, identifying time-domain structural parameters can be considered essentially an optimization problem, wherein the objective is to find the best parameter estimate $\hat{\theta}$ to minimize the error between the measured response $\mathbf{y}(t_k)$ and the predicted response $\hat{\mathbf{y}}(t_k)$ [13, 14]. The objective function is defined as the sum of the least square errors between $\hat{\mathbf{y}}(t_k)$ and $\mathbf{y}(t_k)$

$$\Pi\left(\theta\right) = \sum_{k=1}^{n_t} \mathbf{r}_k^T\left(\theta\right)\mathbf{r}_k\left(\theta\right) \tag{6.1}$$

where $\mathbf{r}_k(\theta)$ is the residue vector, given by

$$\mathbf{r}_k\left(\theta\right) = \hat{\mathbf{y}}\left(t_k\right) - \mathbf{y}\left(t_k\right) \tag{6.2}$$

where t_k is the instant in time, with $k = 0, 1, 2, \cdots, n_t$ and n_t denoting the total number of sampling points.

As shown in Figure 6.1, consider that $\mathbf{F}(t_k)$ and $\mathbf{y}(t_k)$ are the unknown excitation input and measured response output vectors, respectively, of the structural system. The response output $\hat{\mathbf{y}}(t_k)$ of the system can be predicted using the following analytical model:

$$\hat{\mathbf{y}}\left(t_k\right) = f\left(\hat{\mathbf{F}}\left(t_k\right), \hat{\theta}\right) \tag{6.3}$$

where $\hat{\mathbf{F}}(t_k)$ denotes the estimated input vector for the case of output-only identification where $\mathbf{F}(t)$ is unavailable; and $\hat{\theta}$ represents an estimated set of the unknown structural parameters $\theta = \left\{\theta_1, \theta_2, \cdots, \theta_{n_\theta}\right\}^T \in \mathbb{R}^{n_\theta}$, where n_θ is the number of parameters.

Finally, system identification can be summarized as the process to solve the following nonlinear optimization problem in order to determine the optimal parameters:

$$\hat{\theta} = \arg\min_{\theta^l \le \theta \le \theta^u} \left\{\Pi\left(\theta\right)\right\} \tag{6.4}$$

where θ^l and θ^u are the lower- and upper-bound vectors, respectively, of the parameters to be identified. In this study, Eq. (6.4) is solved using the damped Gauss-Newton algorithm.

6.1.2 Force Identification Based on Structural Displacement Measurements

As shown in Eq. (6.3), the forward analysis is required to obtain predicted system responses. Therefore, in order to complete the input–output analysis, the unknown excitation $\mathbf{F}(t)$ should first be estimated.

The equation of motion of a structure subjected to external forces can be written as

$$\mathbf{M}\ddot{\mathbf{x}}(t)+\mathbf{C}\dot{\mathbf{x}}(t)+\mathbf{K}\mathbf{x}(t)=\mathbf{L}\mathbf{F}(t) \tag{6.5}$$

where $\mathbf{M}$, $\mathbf{C}$, and $\mathbf{K}$ are the system mass, damping, and stiffness matrices; $\ddot{\mathbf{x}}(t)$, $\dot{\mathbf{x}}(t)$, and $\mathbf{x}(t)$ are the acceleration, velocity, and displacement vectors, respectively, of the system under external excitation $\mathbf{F}(t)$; and $\mathbf{L}$ is the global load transformation matrix. In this study, viscous damping is assumed, and Rayleigh damping is adopted to model the energy-dissipation mechanism: $\mathbf{C} = \alpha\mathbf{M}+\beta\mathbf{K}$. The constants α and β are obtained from $\alpha = 2\varsigma\omega_1\omega_2/(\omega_1+\omega_2)$ and $\beta = 2\varsigma/(\omega_1+\omega_2)$, where ς is the damping ratio and ω_1 and ω_2 are the first two natural frequencies.

The state-space representation of Eq. (6.5) can be written as

$$\dot{\mathbf{z}}(t)=\mathbf{A}_c\mathbf{z}(t)+\mathbf{B}_c\mathbf{F}(t) \tag{6.6}$$

with

$$\mathbf{z}(t)=\begin{bmatrix}\mathbf{x}(t)\\\dot{\mathbf{x}}(t)\end{bmatrix},\mathbf{A}_c=\begin{bmatrix}\mathbf{0}&\mathbf{I}\\-\mathbf{M}^{-1}\mathbf{K}&-\mathbf{M}^{-1}\mathbf{C}\end{bmatrix}\text{ and }\mathbf{B}_c=\begin{bmatrix}\mathbf{0}\\\mathbf{M}^{-1}\mathbf{L}\end{bmatrix} \tag{6.7}$$

where $\mathbf{A}_c$ is the continuous system matrix, $\mathbf{B}_c$ is the input matrix, and $\mathbf{I}$ is an identity matrix. When displacement measurements are available at a certain limited number of nodes of a structure, the output vector can be formulated as $\mathbf{y}(t) = \mathbf{R}\mathbf{x}(t)$, with $\mathbf{R}$ being the output influence matrix, which depends on the sensor location information. Consequently, the measurement can be expressed as

$$\mathbf{y}(t)=\mathbf{C}_c\mathbf{z}(t) \tag{6.8}$$

where $\mathbf{C}_c = [\mathbf{R}\ \mathbf{0}]$ is the continuous output matrix.

Discretizing Eqs. (6.6) and (6.8) yields

$$\mathbf{z}(k+1)=\mathbf{A}_d\mathbf{z}(k)+\mathbf{B}_d\mathbf{F}(k) \tag{6.9}$$

$$\mathbf{y}(k)=\mathbf{C}_d\mathbf{z}(k) \tag{6.10}$$

where $\mathbf{z}(k)$, $\mathbf{y}(k)$ and $\mathbf{F}(k)$ are discrete vectors at time step $t = k\Delta t$ ($k = 0, 1, 2, \cdots,$ n_t). Here, $\mathbf{A}_d$, $\mathbf{B}_d$, and $\mathbf{C}_d$ are the discrete system state-space matrices given by

$$\mathbf{A}_d = \exp\left(\mathbf{A}_c \Delta t\right), \mathbf{B}_d = \mathbf{A}_c^{-1}\left(\mathbf{A}_d - \mathbf{I}\right)\mathbf{B}_c, \text{ and } \mathbf{C}_d = \mathbf{C}_c \tag{6.11}$$

The substitution of Eq. (6.9) into (6.10) to solve for $\mathbf{y}(k)$ and applying zero initial conditions yields

$$\mathbf{Y} = \mathbf{H}\mathbf{F} \tag{6.12}$$

with

$$\mathbf{Y} = \left\{\mathbf{y}(1); \mathbf{y}(2); \cdots; \mathbf{y}(n_t - 1); \mathbf{y}(n_t)\right\} \tag{6.13}$$

$$\mathbf{F} = \left\{\mathbf{F}(0); \mathbf{F}(1); \mathbf{F}(2); \cdots; \mathbf{F}(n_t - 1)\right\} \tag{6.14}$$

$$\mathbf{H} = \begin{bmatrix} \mathbf{C}_d\mathbf{B}_d & \mathbf{0} & \mathbf{0} & \cdots & \mathbf{0} \\ \mathbf{C}_d\mathbf{A}_d\mathbf{B}_d & \mathbf{C}_d\mathbf{B}_d & \mathbf{0} & \cdots & \mathbf{0} \\ \mathbf{C}_d\mathbf{A}_d^2\mathbf{B}_d & \mathbf{C}_d\mathbf{A}_d\mathbf{B}_d & \mathbf{C}_d\mathbf{B}_d & \cdots & \mathbf{0} \\ \vdots & \vdots & \vdots & \ddots & \vdots \\ \mathbf{C}_d\mathbf{A}_d^{n_t-1}\mathbf{B}_d & \mathbf{C}_d\mathbf{A}_d^{n_t-2}\mathbf{B}_d & \cdots & \mathbf{C}_d\mathbf{A}_d\mathbf{B}_d & \mathbf{C}_d\mathbf{B}_d \end{bmatrix} \tag{6.15}$$

where $\mathbf{Y} \in \mathbb{R}^{n_s(n_t+1)}$ is the assembled measured displacement vector, $\mathbf{F} \in \mathbb{R}^{n_f(n_t+1)}$ is the assembled unknown force vector, and $\mathbf{H} \in \mathbb{R}^{n_s(n_t+1) \times n_f(n_t+1)}$ is the lower block triangular Hankel matrix consisting of the system Markov parameters. The size of matrix $\mathbf{H}$ is determined by the total number of sensors (n_s), the number of data points (n_t), and the number of load points (n_f).

From Eq. (6.12), provided that the system parameters in $\mathbf{H}$ can be estimated from updated structural parameters $\hat{\theta}$, the unknown external force vector $\mathbf{F}$ can be determined from the measured $\mathbf{Y}$ using the ordinary least squares solution [56]:

$$\hat{\mathbf{F}} = \left(\mathbf{H}^{\mathbf{T}}\mathbf{H}\right)^{-1}\mathbf{H}^{\mathbf{T}}\mathbf{Y} \tag{6.16}$$

Note that when $\mathbf{H}$ is singular, the ordinary least squares solution would lead to unbounded solutions. In order to provide a bounded solution, the damped least square approach, well-known as the Tikhonov regularization, should be adopted [13].

6.1.3 Simultaneous Identification Procedure

Figure 6.2 shows the outline of the procedure for simultaneously identifying structural parameters and input forces based on structural response output only. As an output-only inverse identification problem, the estimation of unknown external loads is incorporated in the framework of an iterative parametric optimization process [56]. A finite element model is required to obtain predicted structural responses to compare with the measured ones. The structural parameters as well as the unknown excitation loads can be identified by minimizing the discrepancies between the predicted and measured structural responses, as defined in the objective function in Eq. (6.4), until the convergence criteria are met. The identification problem here is solved by a damped Gauss-Newton optimization algorithm, described in Algorithm 6.1.

The following assumptions are made for the identification procedure: (i) the loading positions, which are used to determine the global load transformation matrix $\mathbf{L}$, are assumed to be known and time-invariant. In comparison, for moving vehicle load identification for coupled vehicle-bridge interaction problems in the literature [13], matrix $\mathbf{L}(t)$ is time-varying at each time step, depending on the axle locations. This assumption may prevent the application of this method to structures subjected to ambient vibration (including combined ambient and wind excitations) unless equivalent loading positions can be predetermined. (ii) In order to avoid non-uniqueness of the solution, the structural mass distribution,

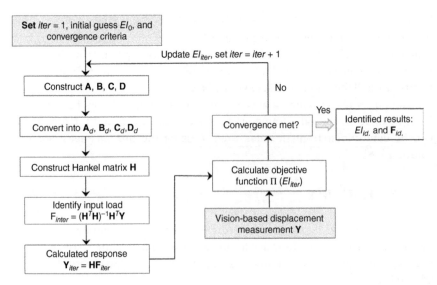

Figure 6.2 Output-only time-domain identification procedure. *Source:* Reproduced with permission of Elsevier.

i.e. the mass matrix $\mathbf{M}$, is assumed to be known a priori. In reality, the structural mass distribution could be obtained from the material types and geometries in the design drawings, with fewer uncertainties compared to other structural parameters, such as stiffness, so this assumption is considered reasonable. (iii) Viscous damping is assumed, and the damping ratio ς should be estimated from the structural material and the measured displacement responses before starting the identification procedure. More discussion of damping is presented in Section 6.2.3.

Algorithm 6.1   Iterative Procedure for the Simultaneous Identification Algorithm

---

Given the measurement vector $\mathbf{Y}$;

Choose initial guess $\boldsymbol{\theta}^{(0)}$ and select the damping factor $h^{(0)}$ and the convergence criteria, i.e. the relative error of the objective function $\Pi(\boldsymbol{\theta})$ is smaller than a tolerance value Δ (e.g. $\Delta = 1 \times 10^{-6}$);

  Set the iteration number index $iter \leftarrow 0$;

While $(|\Pi(\boldsymbol{\theta}^{iter+1}) - \Pi(\boldsymbol{\theta}^{iter})|/|\Pi(\boldsymbol{\theta}^{iter})| > \Delta)$ **do**

  State-space representation of the system using $\boldsymbol{\theta}^{(iter)}$, construct $\mathbf{A}$, $\mathbf{B}$, $\mathbf{C}$, $\mathbf{D}$ and the Hankel matrix $\mathbf{H}_{iter}$;

  Obtain the predicted moving forces: $\mathbf{F}_{iter} = \hat{\mathbf{F}}_k$;

  Calculate the predicted system response: $\mathbf{Y}_{iter} = \mathbf{H}_{iter}\mathbf{F}_{iter}$;

  Calculate the objective function: $\Pi(\boldsymbol{\theta}^{iter})$;

  Calculate the Jacobian matrix of the residual function:

$$
\left[\mathbf{J}_{\mathbf{r}} \right]_{ij} = \frac{\partial \mathbf{r}_i \left(\boldsymbol{\theta}^{iter} \right)}{\partial \theta_j}
$$
$$
\approx \frac{\mathbf{r}_i \left(\theta_1, \ldots, \theta_{j-1}, \theta_j + \delta\theta_j, \theta_{j+1}, \ldots, \theta_{n_\theta} \right) - \mathbf{r}_i \left(\theta_1, \ldots, \theta_{j-1}, \theta_j - \delta\theta_j, \theta_{j+1}, \ldots, \theta_{n_\theta} \right)}{2\delta\theta_j}
$$

  Update the parameter set:

$$
\boldsymbol{\theta}^{(iter+1)} = \boldsymbol{\theta}^{(iter)} + \left[\mathbf{J}_{\mathbf{r}}^T \mathbf{J}_{\mathbf{r}} + h^{(iter)} \operatorname{diag}\left(\mathbf{J}_{\mathbf{r}}^T \mathbf{J}_{\mathbf{r}} \right) \right]^{-1} \mathbf{J}_{\mathbf{r}}^T \mathbf{r}_i \left(\boldsymbol{\theta}^{(iter)} \right)
$$

  Update the numerical damping factor $h^{(iter)}$ using the adaptive tracking approach;

  Update the iteration index $iter \leftarrow iter + 1$;

  end while

Specifically, the unknown parameters (as shown in Figure 6.1) that need to be identified are: (i) the external excitation forces $\mathbf{F}(t)$ and (ii) the uncertain structural parameters (here, flexural rigidity EI, which is used to derive the structural stiffness matrix $\mathbf{K}$). The known parameters are: (i) the bridge geometry and material density and (ii) the impact excitation position. The known observations are vision-based displacement measurements at certain nodes.

6.2 Numerical Example

As shown in Figure 6.3, a numerical study of a single-span simply supported bridge subject to impact excitations is presented to demonstrate the applicability and effectiveness of the presented method for identifying structural stiffness as well as input forces. Table 6.1 lists the structural geometric and material parameters, which are adopted to approximate a real-world bridge. The fundamental frequency of the bridge is 2.59 Hz. In this case, the bridge mass and damping ratio are assumed to be known. The simply supported bridge has a constant global stiffness, which can be represented by the product of the beam's elastic modulus E and the moment of inertia I. The reference value of the global stiffness is $EI = 2.3 \times 10^{10} Nm^2$. As shown in Figure 6.3, the bridge is discretized into 36 Euler-Bernoulli beam elements with 74 degrees of freedom without considering axial deformation. In this numerical example, an external impact force is applied at node 5. The global bridge stiffness in Table 6.1 and the true external impact force are used to numerically simulate the bridge displacement measurements. The time step is 0.01 seconds in the simulation.

Studies are carried out to investigate the accuracy, robustness, and convergence of the identification results given different noise levels, number of sensors, and initial estimates of structural parameters. Table 6.2 summarizes all the simulation cases, including three sets of sensor arrangements (from two-sensor to seven-sensor cases)

Figure 6.3 Numerical example.

Table 6.1 Parameters for the numerical example.

Length (m)	Density (kg/m)	Global stiffness (Nm²)	Damping ratio
36	5000	2.3×10^{10}	0.02 for all modes

Table 6.2 Simulation cases.

No. of sensors	Sensor locations	Measurement noise levels (%)
2	18, 19	0, 2, 5, 10
3	18, 19, 20	0, 2, 5, 10
7	16, 17, 18, 19, 20, 21, 22	0, 2, 5, 10

and four measurement noise levels (from 0 to 10%). Figure 6.3 shows the node numbers of the impact and the sensor locations.

The sensor locations are purposely concentrated on the mid-span region of the bridge, in order to simulate a practical setup for a computer vision sensor in which only one camera is used and its field of view is limited to guarantee good measurement resolution. Including the entire bridge in the field of view would reduce the displacement resolution at each measurement point.

To account for measurement noise, Gaussian random noise with zero mean is added to the simulated displacement responses, whose standard deviation is related to the noise level. In this study, four noise levels are studied: 0, 2, 5, and 10%. To minimize uncertainty in the identification results due to random noise, 10 independent Monte Carlo simulations are carried out for each of the noise levels, the mean values of which are adopted as the final identification results.

In the following optimization process, for the damped Gauss-Newton optimization algorithm, the damping ratio is set to 0.05, and the convergence criterion is either that the relative error of the objective function $\Pi(\theta)$ is smaller than a tolerance value ($\Delta = 1 \times 10^{-10}$) or the number of iterations reaches the user-defined maximum value 100. According to the procedure in Figure 6.2, the structural stiffness can be updated iteratively and the external excitation $\mathbf{F}$ estimated accordingly until the convergence criterion is met.

6.2.1 Robustness to Noise and Number of Sensors

Here, the initial value of the global bridge stiffness is set to $0.8EI = 1.84 \times 10^{10} \text{Nm}^2$. The lower and upper bounds are set to $0.1EI$ and $10EI$, respectively. Figure 6.4 plots the evolutionary process of the updated bridge stiffness for the three sensor arrangements and four noise levels from a typical Monte Carlo analysis. It is seen that the stiffness converges to or close to the true value $2.30 \times 10^{10} \text{Nm}^2$ quickly for all the scenarios.

To further quantify and better visualize identification errors, Figure 6.5 compares the relative errors of the identified bridge stiffness with each column representing the average identification error for 10 independent Monte

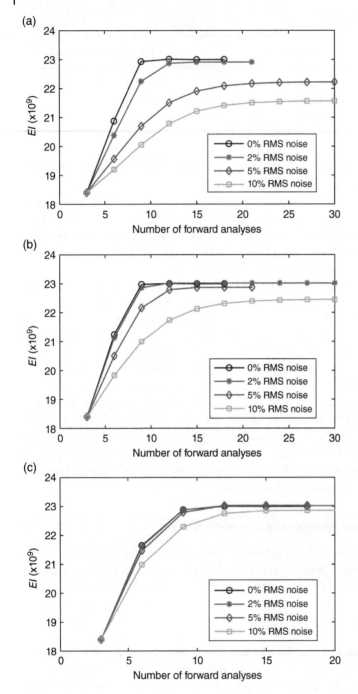

Figure 6.4 Effect of the number of sensors and noise level on the evolution of bridge stiffness identification: (a) two sensors; (b) three sensors; (c) seven sensors. *Source:* Reproduced with permission of Elsevier.

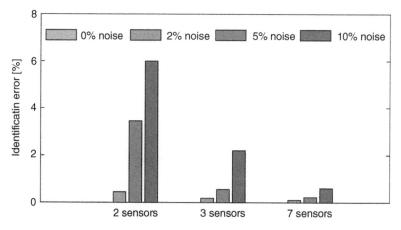

Figure 6.5 Identification errors for bridge stiffness. *Source:* Reproduced with permission of Elsevier.

Carlo analyses. It is observed that the relative error of the identified global stiffness is zero for all three sensor cases when the noise level is 0%. The identification error increases as the noise level increases. The error decreases as the number of sensors increases. For the 10% noise level, the maximum error of the identified stiffness is 6.2% for the two-sensor scenario and only 0.6% for the seven-sensor scenario. Therefore, it can be concluded that the output-only simultaneous identification method is robust to measurement noise, and the robustness can be further enhanced by increasing the displacement measurement points. In this regard, the computer vision sensor has a significant advantage over the conventional method because one camera can simultaneously measure multiple points.

Take the three-sensor case, for example. Figure 6.6 compares the identified external impact forces from displacement measurements at nodes 18, 19, and 20 for the four noise levels. Furthermore, Figure 6.7 compares the predicted displacement responses subject to the identified impact forces with the reference (i.e. measured) response at node 19 given the four levels of measurement noise. In Figure 6.6a, the identified impact force time history matches the reference one perfectly when the noise level is 0%, and so does the displacement time history in Figure 6.7a. For the noisy cases, there are fluctuations in the identified impact force time histories, especially when the measurement noise reaches 10%. However, the predicted displacement time histories under the noisy input forces match the measured ones well.

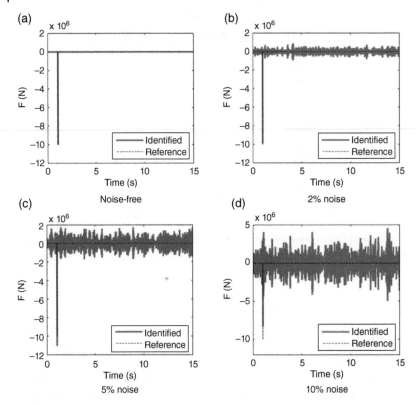

Figure 6.6 Comparison of identified and reference impact forces considering noise: (a) noise-free; (b) 2% noise; (c) 5% noise; (d) 10% noise. *Source:* Reproduced with permission of Elsevier.

6.2.2 Robustness to Initial Stiffness Values

The accuracy of the identification results also depends on the initial values of the bridge stiffness. Figure 6.8 depicts the evolution process of the identified stiffness starting from four different initial stiffness values, including 0.6, 0.8, 1.3, and 1.5 times the true *EI* values for the three-sensor cases at four levels of measurement noise. It is observed that all the parameters efficiently converge to or close to the true value $2.30 \times 10^{10} \mathrm{Nm}^2$, although a higher noise level caused more iterations to converge. Therefore, this simultaneous identification method is also robust to initial stiffness values.

6.2.3 Robustness to Damping Ratio Values

As mentioned earlier, prior to the identification process, the damping ratio should first be estimated from measured displacement responses by using the logarithmic decrement method or other analysis techniques. However, in practical

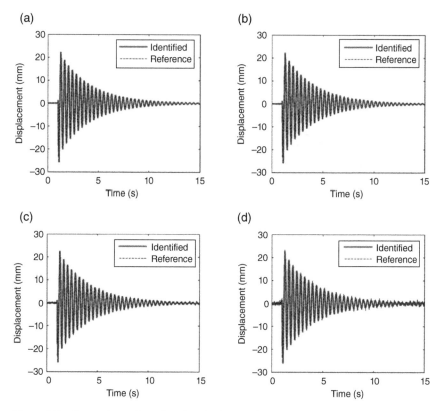

Figure 6.7 Comparison of predicted and reference/measured displacement responses at node 19: (a) noise-free; (b) 2% noise; (c) 5% noise; (d) 10% noise.

applications, damping estimation errors may arise from such uncertainties as the selected number of periods and measurement errors [56]. To construct the damping matrix **C** in Eq. (6.5), damping ratios of two selected modes are needed to obtain the two coefficients in the Rayleigh damping model. Wavelet transforms can be used to uncouple the vibration modes before applying the logarithmic decrement method to estimate the damping ratio for each of the vibration modes.

For this numerical example, to simulate the "measured" displacement responses, a true damping ratio of 0.02 is assumed for all modes. In order to investigate the effects of damping estimation errors on the identification results, the damping ratio is varied among $\varsigma = 0.005$, 0.01, and 0.1, and the resulting identification results are compared with those using the true damping value $\varsigma = 0.02$. Take the three-sensor 2% noise case, for example: for an initial value of $0.8EI = 1.84 \times 10^{10} \text{Nm}^2$, the stiffness quickly converges to the true value of $2.30 \times 10^{10} \text{Nm}^2$, as shown in Figure 6.9. However, as shown in Figure 6.10, for the 2% noise case, there are fluctuations in the identified impact force time histories.

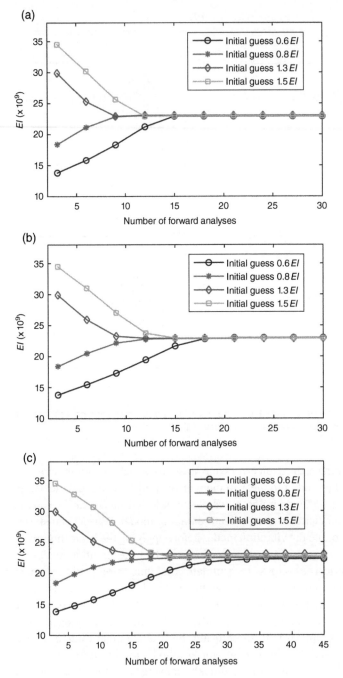

Figure 6.8 Effect of the initial stiffness value on the evolution of bridge stiffness identification: (a) 2% noise; (b) 5% noise; (c) 10% noise.

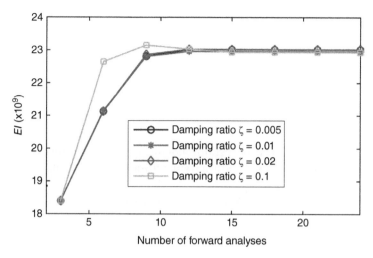

Figure 6.9 Effect of the damping estimate on the evolution of bridge stiffness identification. *Source:* Reproduced with permission of Elsevier.

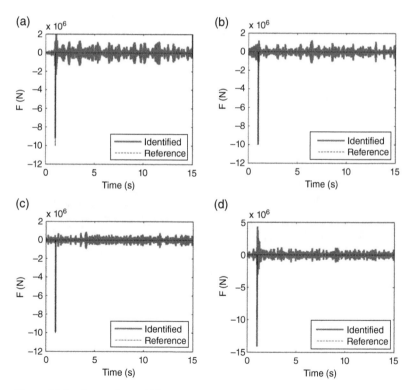

Figure 6.10 Comparison of identified and reference impact forces considering damping error: (a) ς = 0.005; (b) ς = 0.01; (c) ς = 0.02; (d) ς = 0.1. *Source:* Reproduced with permission of Elsevier.

In addition, the identified peak value of the impact force is slightly smaller than the reference (true) one when the damping ratio is underestimated at $\varsigma = 0.005$, and the peak value of the impact force is larger than the reference one when the damping ratio is overestimated at $\varsigma = 0.1$.

6.3 Experimental Validation

To further validate this simultaneous identification method, a laboratory experiment is carried out on a simply supported beam structure at the SMaRT Laboratory at Columbia University. As shown in Figure 6.11, the same 1.6 m scaled bridge model was used to evaluate the vision sensor's performance in Section 3.4 of Chapter 3.

6.3.1 Test Description

The computer vision sensor is used to measure the beam displacement during hammer impact tests. A camera equipped with a zoom lens is fixed on a tripod and placed approximately 8 m from the specimen. The camera and the beam are in the same elevation, i.e. the optical lens axis is perpendicular to the beam's side surface. As shown in Figure 6.12, a series of black dots, numbered 2–31, along the beam's side surface are used as targets for motion tracking. As a reference, the displacements are also measured by two high-accuracy laser displacement sensors at points 9 and 16.

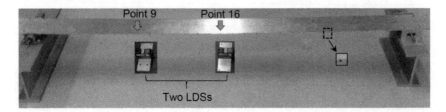

Figure 6.11 Impact test setup.

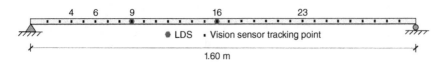

Figure 6.12 Measurement points. *Source:* Reproduced with permission of Elsevier.

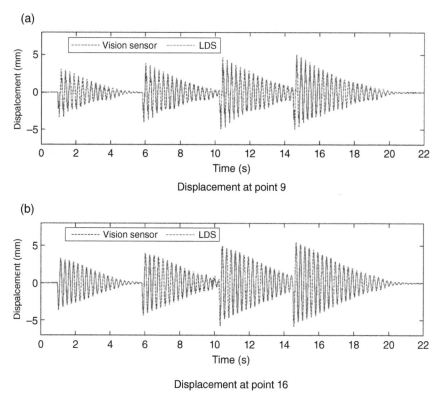

Figure 6.13 Comparison of displacement measurements: (a) displacement at point 9; (b) displacement at point 16. *Source:* Reproduced with permission of Elsevier.

Four repeated hammer impacts are applied at point 4 to excite the beam. The impact force is measured by the force transducer embedded in the hammer. To confirm the measurement accuracy of the vision sensor, the beam displacement time histories measured by the vision sensor at points 9 and 16 are first compared with those from the reference laser displacement sensors (LDSs). As shown in Figure 6.13, excellent agreement can be observed between the displacements measured by the two types of sensors, with normalized root mean squared errors (NRMSEs) of 1.8% at point 9 and 1.2% at point 16.

6.3.2 Identification Results

While the displacements of all 30 target points can be simultaneously measured by the computer vision sensor, only two displacement time histories measured at points 9 and 16, as plotted in Figure 6.13, are used to identify the beam stiffness and the hammer impact forces.

Four initial estimates of the beam stiffness are considered: i.e. $EI_0 = 38, 50, 76,$ 88 Nm^2. Recall that the theoretical value of the beam stiffness is $EI = 63$ Nm^2, as presented in Table 4.3. During optimization, the lower and upper bounds are set to $0.1EI = 6.3$ Nm^2 and $10EI = 630$ Nm^2, respectively.

The damping ratio of the beam is estimated as $\varsigma = 0.013$ from the measured displacement time histories in Figure 6.13 using the logarithmic decrement method. Here, the displacement response is considered to be dominated by the first-mode vibration component, and this damping ratio is assumed to be identical for the second mode. The damping matrix in Eq. (6.5) can then be computed using the Rayleigh damping model based on the initial estimate of the beam stiffness matrix, and updated during the iterative parametric updating process. For example, for initial beam stiffness $EI_0 = 38$ Nm^2, the first two natural frequencies are initially calculated as $\omega_1 = 19.6$ rad/s and $\omega_2 = 78.3$ rad/s through eigenvalue analysis, and the two Rayleigh damping coefficients are $\alpha = 0.3$ and $\beta = 2 \times 10^{-4}$.

As plotted in Figure 6.14, the stiffness converges to the same value after 15 iterations for all 4 initial stiffness values. The identified beam stiffness is 62.4 Nm^2, which is very close to the theoretical value, i.e. $EI = 63$ Nm^2.

The identified time history of the hammer impact force is shown in Figure 6.15. The four repeated hammer impact peaks, clearly observable, match the measured peaks. There are small fluctuations between the identified peaks, which are comparable to those in Figure 6.6b of 2% measurement noise in the numerical example.

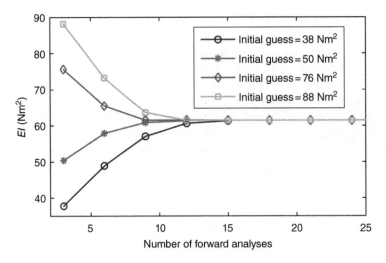

Figure 6.14 Beam stiffness identification from different initial stiffness values. *Source:* Reproduced with permission of Elsevier.

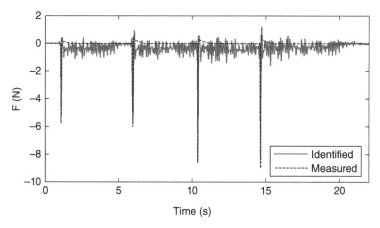

Figure 6.15 Identified and measured hammer impact forces. *Source:* Reproduced with permission of Elsevier.

With the structural parameters identified, the analytical model can be used to predict structural responses at various locations that are not measured or are difficult to measure with sensors but are useful for structural maintenance and management purposes. To demonstrate this application, the displacements at points 6 and 23 are predicted using the analytical model with the simultaneously identified beam stiffness ($62.4\,\text{Nm}^2$) and hammer impact forces (in Figure 6.15). They agree very well with the true time histories measured by the vision sensor, as shown in Figure 6.16. The NRMS error is found to be 1.5% at point 6 and 1.1% at point 23.

Therefore, it can be concluded that noncontact vision-based displacement measurements at two points can be utilized to conveniently and accurately update structural stiffness and identify external impact forces. By increasing the displacement measurement points, this identification method can also detect damage on the beam that is defined as a relative reduction in element stiffness. However, when using one camera to cover a large field of view to include more measurement points, the displacement resolution and accuracy at each measurement point would be reduced.

6.4 Summary

Motivated by the significant advantages of the computer vision sensor, such as the ability to measure noncontact multipoint displacements, low cost, and ease of operation, this chapter presented a novel method for simultaneously identifying both structural parameters and external excitation forces by utilizing vision-based

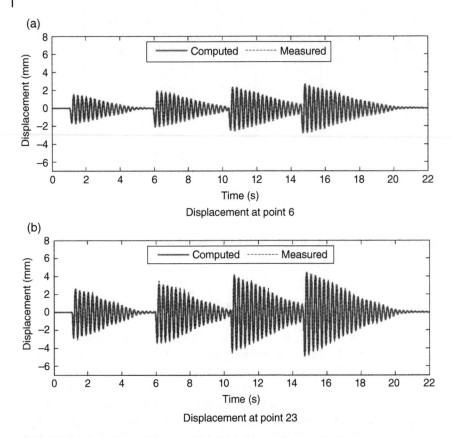

Figure 6.16 Comparison of the predicted and measured beam displacement:
(a) displacement at point 6; (b) displacement at point 23. *Source:* Reproduced
with permission of Elsevier.

displacement measurements. The identification problem was formulated as an
iterative, time-domain output-only optimization process by minimizing discrepancies between measured and computed displacement responses. While updating
structural parameters (such as stiffness), unknown external excitation forces
(such as impact forces) can also be iteratively estimated based on the established
state-space representation of the input–output relationship.

As shown via numerical simulations, bridge stiffness and external impact forces
can be successfully identified from structural displacement time histories measured at a limited number of sensor locations with up to 10% RMS noise level.
Identification accuracy can be further improved by increasing the number of
measurement points. The robustness and fast convergence of the method were
demonstrated via different initial values of the structural parameters. Through an

experimental laboratory test on a beam structure, it was demonstrated that displacements measured by the vision sensor at two points were sufficient for accurately identifying both the beam stiffness and hammer impact forces. Using the updated analytical model and the identified input forces, the structural response at any point could be accurately predicted.

While the examples presented in this chapter were beam structures, the output-only time-domain simultaneous identification method is applicable to other types of structures and other types of excitation forces beyond impact force (such as seismic ground motion). The computer vision sensor makes it easier to implement this method.

To better understand the presented output-only identification procedure, the case of the numerical example (three sensors and 2% noise) in Section 6.2 is solved using MATLAB scripts as follows.

MATLAB Code – Simultaneous Identification of Structural Parameters and Excitation Forces [56]

```
%-------------------------
% Description: Main script
%-------------------------
close all; clear all; clc;
global M n N_t dt YY Xdd R1 R t LL sensor1 meanp EL
bcdof Sys_dof No_nel No_el P
tStart = cputime;


%-------------- Analysis setup
Model_input_info;    % Model material and geometric
                          properties
n=Sys_dof;           % DOFs of beam bridge system
NoiseLevel = 0.02;   % Noise level in the simulated
                          measurements
Dim = 1;             % Number of parameters to be
                          identified
MCrun = 1;           % Monte Carlo simulations
SN=18:1:20;          % Sensor location: 3 sensors at
                          node 18, 19, 20
sensor1=2*SN-1;      % DOF index
sensor=2*SN-1;


%-------------- Generate Hammer impulse excitation
```

```
Fs=100;
dt=1/Fs;
t=0:dt:15;
N_t=length(t);
P=zeros(N_t,1);
n_index=find(t==1);
P(n_index)=-1e7;
node_Hammer=5;                  % Node # of hammer
                                  excitation
dof_hammer=2*node_Hammer-1;     % DOF # of hammer
                                  excitation

for NoiseLevel_index=1:length(NoiseLevel);
%------------ Problem specific variables
Func = 'state';                 % Cost function to be
                                  optimized
ic = 0.8*ones(1, Dim);          % Initial guesses
lb = 0.5*ones(1, Dim);          % Lower bounds
ub = 1.5*ones(1, Dim);          % Upper bounds
%------------ Damped Gauss-Newton Process
global history2
  history2.x = [];
  history2.funccount = [];
hStep = 1e-1;
dXOpt = hStep*ones(1, Dim);
XOptMC = zeros(MCrun, Dim);
for r = 1:MCrun
observed_data;      % Simulate observed/measured data
fprintf(1,['Monte Carlo Runing:  r= ',num2str(r),' \n']);
   Options = optimset('Algorithm', 'trust-
region-reflective', 'Display', 'iter-detailed',
'ScaleProblem', 'Jacobian', 'TolFun', 1e-10,
'FinDiffType', 'central', 'FinDiffRelStep', dXOpt,
'MaxIter', 100, 'TolX', 1e-10, 'UseParallel',
'always', 'OutputFcn', @outfun2);
   [XOpt, ~, ~, ~, ~, ~, ~] = lsqnonlin(Func, ic, lb,
ub, Options);
XOptMC(r, :) = XOpt;
end
   TCPU = cputime - tStart;
disp(['CPU time = ', num2str(TCPU)]);
end
```

```
%----------------------------------------------------------
% Description: Model material and geometric properties
%----------------------------------------------------------
global bcdof No_nel No_dof No_el nodes

LB = 36;                    % length of Bridge (m)
No_el = LB;                 % Number of elements (assign
                              even No. of elements)
EL = LB/No_el;              % Element length
No_node = No_el + 1;        % Number of nodes

No_nel=2;                   % Number of nodes per element
No_dof=2;                   % Number of DOFs per node
Sys_dof=No_node*No_dof;     % Total system DOFs

%-------------- Nodal coordinates
gcoord=[(1:No_node)',(0:EL:LB)',zeros(No_node,1)];
% x, y coordinates in global coordinate system
gcoord=gcoord(:,2:end);

%-------------- Nodal connectivity of the elements
nodes=[ (1:No_el)',(1:No_node-1)',(2:No_node)'];
nodes=nodes(:,2:end);

%-------------- Boundary conditions
ConNode=[ 1,    1,    0;...
No_node,  1,    0];
ConVal =[ 1,    0,    ;...
No_node,  0,    ];

[n1,n2]=size(ConNode);
bcdof=zeros(Sys_dof,1); % Initializing the vector bcdof
bcval=zeros(Sys_dof,1); % Initializing the vector bcval

for ni=1:n1                 % Calculate the constrained DOFs
ki=ConNode(ni,1);
bcdof((ki-1)*No_dof+1:ki*No_dof)=ConNode(ni,2:No_
dof+1);
end
```

```
%------------------------------------------------
% Description: Simulate 'observed/measured' data
%------------------------------------------------
global M n N_t dt YY Xdd R1 R t F1 F2 sensor1 meanp EL
bcdof Sys_dof No_nel No_el

%---------- Generate force vector
LL=zeros(Sys_dof,1);
LL(dof_hammer,:)=1;
Q=LL*P';

%---------- Material parameters
rho = 5000;     % Linear density (kg/m)
EI=2.3e10;      % Flextural rigidity EI (N/m^2)
True1 = EI;
meanp = EI;
True = True1./meanp;
prop(1)=rho;
prop(2)=EI;
prop(3)=EL;

%---------- Assemble global mass and stiffness matrix
kk=zeros(Sys_dof,Sys_dof);        % Initialization of
                                    system stiffness
                                    matrix
mm=zeros(Sys_dof,Sys_dof);        % Initialization of
                                    system mass matrix
index=zeros(No_nel*No_dof,1);     % Initialization of
                                    index vector

for i=1:No_el
[me,ke]=km_undamage(prop);
nd(1)=nodes(i,1);
nd(2)=nodes(i,2);
index=femEldof(nd,No_nel,No_dof);
kk=femAssemble1(kk,ke,index);
mm=femAssemble1(mm,me,index);
end
[K,M]=femApplybc(kk,mm,bcdof);    % Apply constraints
                                    to matrix equation
                                    [kk]{x}1-4=ff
```

```
%---------- Damping matrix: C=a*M+b*K
Freq = sqrt(eig(K,M))/2/pi;
w1 = Freq(3)*2*pi;
w2 = Freq(4)*2*pi;
zeta1 = 0.02;
zeta2 = 0.02;
beta = 2*(w1*zeta1 - w2*zeta2)/(w1^2 - w2^2);
alpha = 2*zeta1*w1 - beta*w1^2;
C = alpha*M + beta*K;


%---------- State-space method to solve dynamic
MXdd(t)+CXd(t)+KX(t)=Q(t)
nsensor1 = length(sensor1);
R1 = zeros(nsensor1, n);
for indR = 1:nsensor1
R1(indR, sensor1(indR)) = 1;
end
nsensor = length(sensor);
R = zeros(nsensor, n);
for indR = 1:nsensor
R(indR, sensor(indR)) = 1;
end

A = [zeros(n), eye(n); -(M\K), -(M\C)];  % A matrix
B = [zeros(n); M\eye(n)];                % B matrix
c = [R zeros(size(R))];
D = 0;                                   % C and D matrix
sys = c2d(ss(A, B, c, D), dt);
[y, ~, x] = lsim(sys, Q, t, zeros(1, 2*n));
Xdd1 = y;                      % Displacement history
X = x(:, 1:n)';                % Displacement history
Xd =  x(:, 1+n:end)';          % Velocity history

%---------- Add noises to simulated responses
[NN, DOF] = size(Xdd1);
Xdd = zeros(NN, DOF);
for i = 1:DOF
    noiseout = randn(NN, 1);
    N_level=NoiseLevel(NoiseLevel_index);
    Xdd(:,i) = Xdd1(:, i) + N_level*std(Xdd1(:,i))/
std(noiseout)*noiseout;
```

```
end
Xdd = Xdd';                              % Noised response
Xdd(:, 1) = zeros(length(sensor), 1);    % Apply zero
initial conditions @ t=0s;
[~, IA] = intersect(sensor, sensor1);
YY = reshape(Xdd(IA, 2:N_t), nsensor1*(N_t-1), 1);


%--------------------------------
% Description: Objective function
%--------------------------------
function ObjVal = state(Colony)

global M n N_t dt YY Xdd  R t LL sensor1 meanp EL
Sys_dof No_nel No_dof No_el nodes P

EI = Colony*meanp(1);
fprintf(1,['   EI=',num2str(Colony),' \n']);
zeta1 = 0.02;   % Damping ratio: zeta1=zeta2=0.02
zeta2 = 0.02;

%---------- Global mass and stiffness matrix
kk=zeros(Sys_dof,Sys_dof);        % Initialization of
                                    system stiffness
                                    matrix
index=zeros(No_nel*No_dof,1);     % Initialization of
                                    index vector

for i=1:No_el
  ke = (EI/(EL^3))*[12    6*EL   -12    6*EL;
       6*EL   4*EL^2 -6*EL 2*EL^2;
       -12    -6*EL   12   -6*EL;
       6*EL   2*EL^2 -6*EL 4*EL^2];    % Element
stiffness matrix
    nd(1)=nodes(i,1);
    nd(2)=nodes(i,2);
    index=femEldof(nd,No_nel,No_dof); %  System DOFs
associated with each element
```

```matlab
    kk=femAssemble1(kk,ke,index);        %   Assembly
global stiffness matrix
end
K=state_femApplybc(kk);                  %   Apply constraints
to matrix equation

%--------- Damping matrix: C=a*M+b*K
Freq = sqrt(eig(K,M))/2/pi;              % Eigenvalue_analysis;
w1 = Freq(3)*2*pi;
w2 = Freq(4)*2*pi;
beta = 2*(w1*zeta1 - w2*zeta2)/(w1^2 - w2^2);
alpha = 2*zeta1*w1 - beta*w1^2;
C = alpha*M + beta*K;


%--------- Force identification
nsensor = length(sensor1);
nForce = size(LL, 2);
A = [zeros(n), eye(n); -(M\K), -(M\C)];
B = [zeros(n)*LL; M\eye(n)*LL];
c = [R zeros(size(R))];
D=0;
sys = c2d(ss(A, B, c, D), dt);
Ad = sys.a;
Bd = sys.b;
Cd = sys.c;
Gamma_cell = cell(N_t-1, 1);
for indG = 1:N_t-1
  if indG == 1
        Mat = Cd*Bd;
  else
        Mat = Cd*Ad^(indG-1)*Bd;
  end
Gamma_cell(indG, :) = [59];
end
Gamma = cell2mat(Gamma_cell);

H0 = cell(1, N_t-1);
for indI = 1:N_t-1
```

```
  H1 = { [zeros(nsensor*(indI-1), nForce);
Gamma(1:end-nsensor*(indI-1), :)] };
  H0(:, indI) = H1;
end
HH = cell2mat(H0);

FF = (HH'*HH)\(HH')*YY;             % Ordinary least
square for force identification
f = reshape(FF, nForce, N_t-1);    % Identified force

%---------- Recalculate responses from identified
forces
Q=LL*f;

A = [zeros(n), eye(n); -(M\K), -(M\C)];    % A matrix
B = [zeros(n); M\eye(n)];                  % B matrix
c = [R zeros(size(R))];
D = 0;
% C and D matrix
sys = c2d(ss(A, B, c, D), dt);
[y, ~, ~] = lsim(sys, Q, t(1:N_t-1), zeros(1, 2*n));
XXdd = y';

S = XXdd - Xdd(:, 1:N_t-1);         % Response residue
[rr, cc] = size(S);
ObjVal = reshape(S', rr*cc, 1);

fprintf(1,['%%%%%%%%%%%%%%%%%%%%%% ObjVal   Norm
',num2str(norm(ObjVal)),' \n']);

%------------------------------------------------------------
% Function state_femApplybc: Apply constraints to
matrix equation [kk]{x}=ff
%------------------------------------------------------------
function kk=state_femApplybc(kk)
global bcdof
ni=length(bcdof);
sdof=size(kk,1);
for ii=1:ni
    if bcdof(ii)==1
```

```
            for j=1:sdof
                kk(ii,j)=0;
                kk(j,ii)=0;
            end
            kk(ii,ii)=1;
        end
end


%-----------------------------------------------
function [index]=femEldof(nd,No_nel,No_dof)
%-----------------------------------------------
%  Purpose: Compute system dofs associated with each
element
%-----------------------------------------------
    k=0;
    for i=1:No_nel
        start = (nd(i)-1)*No_dof;
            for j=1:No_dof
                k=k+1;
                index(k)=start+j;
            end
    end

function [me,ke]=km_undamage(prop)
%-----------------------------------------------------
%  Purpose: Element Mass and Stiffness matrices
%-----------------------------------------------------
rho=prop(1);
EI=prop(2);
EL=prop(3);
% Element mass matrix
me = (rho*EL/2)*[1        0        0 0;
                    0 1/12*EL^2 0 0;
                    0        0     1 0;
                    0        0     0 1/12*EL^2];
% Element stiffness matrix
ke = (EI/(EL^3))*[12      6*EL    -12      6*EL;
                    6*EL  4*EL^2 -6*EL 2*EL^2;
                    -12    -6*EL    12    -6*EL;
                    6*EL  2*EL^2 -6*EL 4*EL^2];
```

```
function [kk]=femAssemble1(kk,k,index)
%-----------------------------------------------------------
%  Purpose: Assembly of element matrices into the
system matrix
%-----------------------------------------------------------
eldof = length(index);
for i=1:eldof
   ii=index(i);
   for j=1:eldof
      jj=index(j);
      kk(ii,jj)=kk(ii,jj)+k(i,j);
   end
end

function [kk,mm]=femApplybc(kk,mm,bcdof)
%-----------------------------------------------------------
%  Purpose:
%     Apply constraints to matrix equation [kk]
{x}={ff}
%     Apply constraints to eigenvalue matrix equation
[kk]{x}=lamda[mm]{x}
%-----------------------------------------------------------
 ni=length(bcdof);
 sdof=size(kk,1);
 for ii=1:ni
   if bcdof(ii)==1
      for j=1:sdof
         kk(ii,j)=0;
         kk(j,ii)=0;
         mm(ii,j)=0;
         mm(j,ii)=0;
      end
      kk(ii,ii)=1;
      mm(ii,ii)=1e10;
   end
 end
```

```matlab
function stop = outfun2(x, optimValues, state)
%---------------------------------------------------------
% Purpose: Store the iterative Gradient-based opt
results
%---------------------------------------------------------
global history2
stop = false;
   switch state
       case 'iter'
           history2.x = [history2.x; x];
           history2.funccount = [history2.funccount;
optimValues.funccount];
   end
end
```

7

Application in Estimating Cable Force

Cables are generally designed to efficiently carry axial loads for cable-supported structures. Typical examples include stay cables in cable-stayed bridges, vertical hanger cables in suspension bridges, and cables in tower masts and large-span roof structures. Accurately determining cable tension force is of great importance for both construction control and overall structural condition assessment during the cable's service life.

Currently, there are three primary techniques for estimating cable tension forces: direct measurement by devices such as hydraulic jacks and load cells, the magnetic method based on magnetic permeability measurements, and the vibration method. Due to its relatively easy implementation, the vibration method is more widely employed in practical applications. This method is based on the correlation between the cable's natural frequencies and tension force. Conventionally, accelerometers are mounted on the cable surface to measure vibrations, from which cable frequencies are extracted. These sensors need to be wired to external power supply and data-acquisition devices. This is generally expensive to implement on structures with a large number of cables due to the cumbersome and time-consuming sensor installation. While advanced wireless sensors can greatly improve the efficiency of measuring cable vibrations, these contact-type sensors still require physical access to structures in order to install them on the cable surface; this is often very difficult, if not impossible, particularly for structures in operation. Hence, noncontact sensor systems are highly desired for measuring cable tension forces and monitoring their health conditions [2].

In this chapter, the computer vision sensor system, enabled by the robust sub-pixel orientation code matching (OCM) algorithm, is applied for noncontact, multi-point measurement of cable vibration displacements by tracking natural features on the cable, without requiring physical access to the cable. Section 7.2

Computer Vision for Structural Dynamics and Health Monitoring, First Edition.
Dongming Feng and Maria Q. Feng.
© 2021 John Wiley & Sons Ltd.
This Work is a co-publication between John Wiley & Sons Ltd and ASME Press.
Companion website: www.wiley.com/go/feng/structuralhealthmonitoring

presents the procedure for estimating cable tension forces based on vibration measurements. Implementation with real cable structures is discussed in Section 7.3 for construction control of the cable-supported roof structure of the Hard Rock Stadium and in Section 7.4 for suspender rope replacement on the Bronx-Whitestone Bridge. Without interrupting the stadium's construction or the bridge's operation, these two case studies demonstrate that the computer vision sensor is an exceptionally suitable and cost-effective means for estimating cable force and monitoring long-term cable health.

7.1 Vision Sensor for Estimating Cable Force

7.1.1 Vibration Method

The vibration method is based on the correlation between the cable's natural frequencies and tension forces. Among the equations for measuring cable tension, formulations from taut string theory and beam theory are widely used. The string theory simplifies the analysis and can be expressed as

$$T = 4\pi^2 m l^2 \left(\frac{f_n}{n} \right)^2 \tag{7.1}$$

where T is the cable tension force, f_n is the nth measured natural frequency, m is the mass density per unit length, and l is the cable length. Equation (7.1) ignores the sag and bending stiffness of the cable and may introduce unacceptable estimation errors. Based on the vibration equation of an axially loaded beam with hinged end boundary conditions, a formulation from beam theory can be shown as

$$T = 4\pi^2 m l^2 \left(\frac{f_n}{n} \right)^2 - \frac{EI}{l^2} (n\pi)^2 \tag{7.2}$$

where EI is the flexural rigidity of the cable. Although the cable bending stiffness is considered, Eq. (7.2) may still lead to significant errors for force estimates of short, stout cables [58].

On the basis of the cable's transverse vibration and using a curve-fitting method, Fang and Wang [58] proposed a practical formula to estimate the cable tension in an explicit form, which is a unified expression integrating string and beam theories:

$$T = 4\pi^2 m l^2 \frac{f_n^2}{\gamma_n^2} - \frac{EI}{l^2} \gamma_n^2 \tag{7.3}$$

with

$$\gamma_n = n\pi + A\psi_n + B\psi_n^2$$

$$\psi_n = \frac{1}{\chi_n\gamma_n} = \sqrt{\frac{EI}{m\omega_n^2 l^4}}$$

$$A = -18.9 + 26.2n + 15.1n^2$$

$$B = \begin{cases} 290 & (n=1) \\ 0 & (n\geq 2) \end{cases}$$

The accuracy of Eq. (7.3) has been demonstrated through a comparison of experimental and numerical results in the literature [58]. Therefore, this formulation is directly adopted to estimate cable tension forces in this study.

7.1.2 Procedure for Vision-Based Cable Tension Estimation

Figure 7.1 outlines the procedure for estimating cable tension force using the OCM-based computer vision sensor, which can be further detailed as follows [2]:

1) *Set up the computer vision system.* Fix the video camera equipped with a zoom lens on a tripod or a rigid stationary structure at a convenient location away from the cables. The camera is connected to a computing unit such as a PC, which has the real-time displacement measurement software installed.

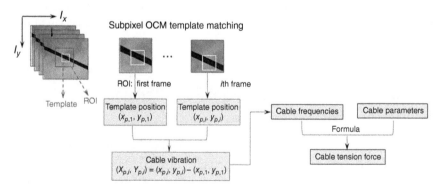

Figure 7.1 Outline of vision-based cable tension measurement. *Source:* Reproduced with permission of Elsevier.

2) *Measure cable vibrations.* Select an initial area on the cable surface as a template to be tracked. To reduce computational time, the search area can be confined to a predefined region of interest (ROI) near the template's location in the previous image. Videos recorded by the camera are digitized into images with specified resolutions and frame rates and streamed into the computer. The measured vibration can then be obtained in real time based on the subpixel OCM algorithm and saved to the computer.

A unique advantage of the computer vision system for this application is that coordinate transformation is unnecessary, since only frequency information is required for estimating cable force, according to Eq. (7.3). In other words, there is no need to determine the scaling factor (units: mm/pixel) to transform the pixel coordinate vibrations into physical coordinate vibrations, which makes the measurement procedure even more efficient and practical.

3) *Perform a Fourier transform, and estimate cable tension.* Perform a Fourier transform to identify the natural frequencies of the cable for the measured vibration time history. Finally, given the cable's geometry and material parameters, estimate the cable tension force according to Eq. (7.3).

Often, it is preferable to shoot a cable from a remote distance. To guarantee the measurement resolution, an optical lens with the proper focal length is needed to zoom in the image and obtain an enlarged cable segment. During measurement, the base of the camera should be fixed and stationary. However, when the vision sensor system has to be placed on the structure itself for the cable measurement, e.g. on the bridge deck, structural vibration-induced camera motion is a major concern. In this case, it is recommended to conduct an additional test by focusing the camera on a stationary object: for example, a segment of the stiff bridge tower above the deck. Assuming that the object is not moving, the measured vibration can be considered camera vibration. The camera vibration frequencies can then be identified and filtered out from the power spectrum of the cable vibration measured by the vision sensor.

7.2 Implementation in the Hard Rock Stadium Renovation Project

During the construction of cable-supported structures, various measurements, such as elevation, deformation, and structural stress, should be taken to ensure safety and quality. Cable tension force is one of the most important indices as it reflects overall structural performance and safety. Conventionally, design cable forces are predicted by analyzing an erection simulation. Due to uncertainties in geometric and material properties as well as boundary conditions used in the simulation model, the predicted cable forces should be updated based on field measurements. As construction proceeds, the cable forces change significantly and need to be closely monitored. Currently, when estimating cable forces with the vibration method, accelerometers are primarily used, and the time-consuming setup delays construction. Therefore, accurate and reliable cable force measurements using the noncontact vision-based sensor from a remote distance is a promising alternative, benefiting both contractors and designers by allowing for construction validation and safety monitoring without delaying construction [2].

7.2.1 Hard Rock Stadium

Thornton Tomasetti, Inc. provided structural engineering services for major renovations at Hard Rock Stadium in Florida, home to the Miami Dolphins NFL team, to upgrade it into a world-class stadium, including the construction of a new long-span, cable-supported canopy that covers the entire seating bowl and protects fans from the elements. The roof framing is structural steel incorporating a series of wide box trusses and deep tied arches. As shown in Figure 7.2, it is supported by a network of cables at each corner, including four backstay tie down (TD) cables, one lower forestay (LF) cable, one upper forestay (UF) cable, one end zone lower backstay (EZLB) cable, one sideline lower backstay (SLLB) cable, one end zone upper backstay (EZUB) cable, one sideline upper backstay (SLUB) cable, one sideline forestay (SLF) cable, one end zone forestay (EZF) cable, two side backstay cables, and two lower hurricane stay cables. At various stages of constructing the

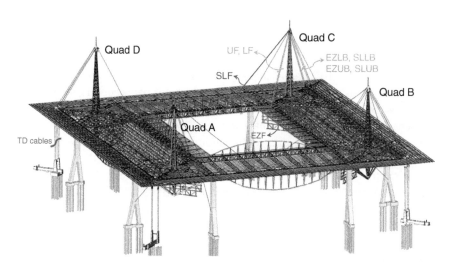

Figure 7.2 Hard Rock Stadium.

roof framing box trusses, the cables at Quad A through Quad D are tensioned and adjusted according to predesigned erection sequences. For each cable, the main activities on site include: uncoiling cable, assembling socket and installing cable, top socket and bottom socket, and tensioning cable.

During construction, the cable forces change significantly as erection proceeds. Therefore, cable forces should be closely monitored, as this is one of the most important indices for construction control. To ensure that the cable forces reach their design values, the computer vision sensor is used to monitor them throughout the roof erection process. For simplicity, measurement results from 24 selected cables are presented and discussed in this section, including the 16 TD cables at the four quads and the 8 inclined cables (LF, EZLB, SLLB, UF, EZUB, SLUB, SLF, and EZF) at Quad C. The inner TD cables at each quad are denoted TD_A and TD_B, and the outer TD cables TD_C and TD_D, as shown in Figure 7.2.

The cable lengths are tabulated in Table 7.1. All the cables have the same diameter (0.128 m), are made of the same steel material with elastic modulus 163 GPa, and have mass density 90.72 kg/m. It is still a challenging task to accurately determine the effective vibration length of a cable, mainly due to uncertainties related to boundary conditions from constraints and anchorage (socket) systems at both ends of the cables. Although the cables are pin-connected at both ends, as shown in Figure 7.3, in real structures, the cable ends are often restrained from rotating due to significant friction when cables are under tension from a dead load. Therefore, in this study, even with pin connections, cable vibrations due to ambient or impact excitations are characterized by assuming a fixed connection to the adjacent structures.

7.2.2 Test Description

For the vision sensor system shown in Figure 7.4, the adopted video camera (Point Grey, FL3-U3-13Y3M-C) has a CMOS-type sensor with a maximum resolution of 1280×1024 pixels and a maximum rate of 150 frames per second (fps), and the focal length of the optical lens varies in the range 16–160 mm with manual focus. The sensor uses a global shutter that captures the entire image frame at the same instant. This is in contrast to a rolling shutter in which each image frame is recorded by scanning pixels row by row or column by column. In other words, the rolling shutter method may produce distortions when recording fast-moving objects, the effect of which on the measurement results should be rectified when this type of video camera is used. Given that tensioned cables in most civil infrastructure have a fundamental frequency under 10 Hz, here, a sampling rate of 50 fps is adopted, which is capable of measuring vibrations up to 25 Hz. This is considered sufficient to capture the dominant frequency components of the cable vibration.

Table 7.1 Cable length, measured cable frequencies, and tension discrepancies w.r.t. reference values.

	Quad A				Quad B				Quad D			
	TD_A	TD_B	TD_C	TD_D	TD_A	TD_B	TD_C	TD_D	TD_A	TD_B	TD_C	TD_D
Cable length (m)	34.76	34.76	35.22	35.22	34.76	34.76	35.22	35.22	34.76	34.76	35.22	35.22
First freq. (Hz)	1.89	1.89	1.90	1.86	2.46	2.46	2.49	2.46	2.63	2.69	2.65	2.58
Tension discrepancy (%)	−5.6	−5.6	−1.1	−5.6	−0.2	−0.2	3.2	1.1	−2.0	2.6	2.4	−3.3

	TD_A	TD_B	TD_C	TD_D	LF	EZLB	SLLB	UF	EZUB	SLUB	EZF	SLF
Cable length (m)	34.76	34.76	35.22	35.22	55.81	58.65	58.65	57.78	60.53	60.53	79.42	90.20
First freq. (Hz)	2.81	2.78	2.69	2.75	2.05	2.02	1.96	2.02	1.93	1.87	1.08	2.02
Tension discrepancy (%)	5.6	3.3	−0.4	4.2	−0.4	0.1	−3.8	2.0	4.3	0.1	2.5	0.9

Figure 7.3 Typical cable assembly.

(a)

(b)

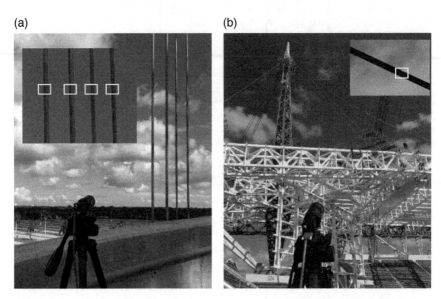

Figure 7.4 Implementation of the computer vision sensor in Hard Rock Stadium: (a) tie down cables; (b) inclined cables. *Source:* Reproduced with permission of Elsevier.

The vibrations of the four TD cables at each quad of the stadium are simultaneously measured using one camera. The approximately mid-span cable segments framed by squares in Figure 7.4 are registered as templates for motion tracking. As shown in this figure, it would be highly difficult and dangerous to install conventional contact-type sensors such as accelerometers on the surface of these EZF cables. The camera is set on the stadium seating bowl approximately 200 m away to remotely track cable vibrations, completely eliminating the need to access the cables.

7.2.3 Estimating and Validating Cable Force

Ambient vibrations are measured for long, inclined cables. For the short TD cables at the four quads, manually induced free vibrations are measured, due to their small ambient vibration in the absence of strong wind. Figures 7.5–7.8 plot the displacement time histories, together with their PSD amplitudes, measured by the computer vision system for one TD cable at each quad. Table 7.1 tabulates the first

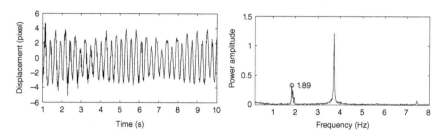

Figure 7.5 Measured vibration and PSD function of TD_A cable at Quad A.

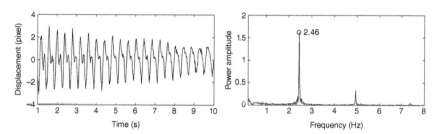

Figure 7.6 Measured vibration and PSD function of TD_B cable at Quad B.

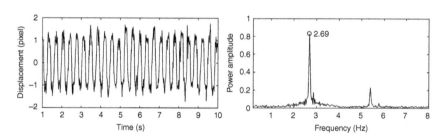

Figure 7.7 Measured vibration and PSD function of TD_C cable at Quad C.

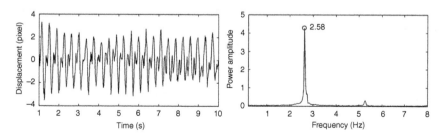

Figure 7.8 Measured vibration and PSD function of TD_D cable at Quad D.

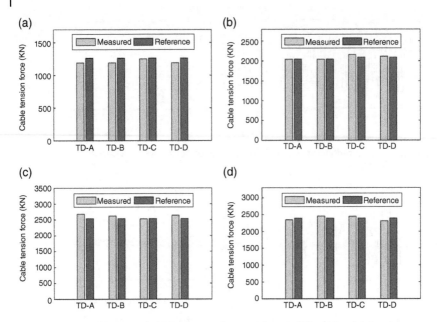

Figure 7.9 Measured tension forces vs. reference forces for TD cables: (a) Quad A; (b) Quad B; (c) Quad C; (d) Quad D.

natural frequency of each of the TD cables at the four quads, identified from the PSD amplitudes.

Using the measured first natural frequency, the tension force is estimated for each of the cables according to Eq. (7.3). In order to validate these estimates, the cable forces are directly measured using jacking load cells as reference values. Figure 7.9 shows the tension forces estimated from the computer vision sensor measurement (referred to as Measured) and those directly measured by the load cells (referred to as Reference). Their discrepancies are further quantified, as shown in Table 7.1. It is observed that the cable forces measured by the vision sensor agree well with those from the reference load cells, with a maximum difference of 5.6%.

As mentioned, ambient vibrations are measured for the long, inclined cables. Figures 7.10–7.13 plot the measured vibration time histories of cables SLLB, EZUB, EZF, and SLF, together with their PSD amplitudes. The measured first natural frequencies of the eight inclined cables at Quad C are listed in Table 7.1. Figure 7.14 compares the estimated cable forces based on a vision sensor with the reference values, the discrepancies of which are tabulated in Figure 7.13, with a maximum difference of 4.3% for the EZUB cable.

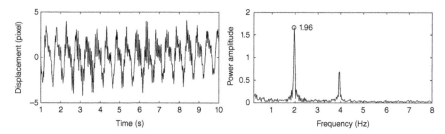

Figure 7.10   Measured vibration and PSD function of SLLB cable at Quad C.

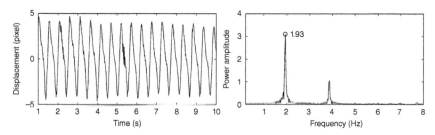

Figure 7.11   Measured vibration and PSD function of EZUB cable at Quad C.

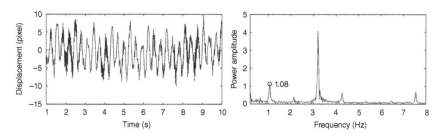

Figure 7.12   Measured vibration and PSD function of EZF cable at Quad C.

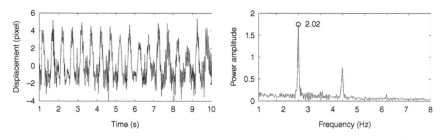

Figure 7.13   Measured vibration and PSD function of SLF cable at Quad C.

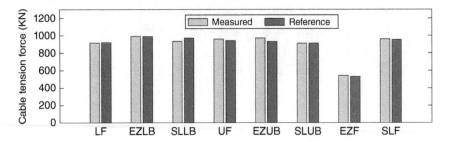

Figure 7.14 Measured tension forces for inclined cables using the vision sensor vs. reference values.

Therefore, the presented method based on cable vibrations measured by the noncontact vision sensor can accurately and conveniently determine cable tension forces. Note that for the cable force measurements, the small discrepancies may be caused by the following: (i) inherent error when using Eq. (7.3) to estimate cable force; (ii) uncertainties in determining the effective cable vibration length; and (iii) measurements by the vision sensor being carried out a few days after all the TD cables were tensioned by jacking devices and measured by the reference load cells. Ongoing work to erect the roof during the process may change the actual cable forces slightly.

As an illustration, the following MATLAB commands are used to obtain cable vibration frequencies and calculate cable tension forces based on the vibration method.

MATLAB Code – Vibration Method for Estimating Cable Tension Force

```
close all; clear; clc
%-------------- Loading measured cable displacement
by vision-based sensor
data=csvread('data.csv');
t=data(:,1);
dt=t(2)-t(1);
Fs=1/dt;
Dy=data(:,2);
NT=length(Dy);

figure(1)
plot(t,Dy)            % Plot measured cable displacement
xlabel('Time[s]', 'FontName','Arial','fontsize', 14);
ylabel('Displacement [pixel]', 'FontName','Arial',
'fontsize', 14);
```

```
title('Measured Displacement by Vision Sensor','Font
Name','Arial','fontsize', 14);
set(gca,'FontName','Arial', 'fontsize', 14);
set(gcf,'Position',[100 200 800 600]);

%--------------- FFT of measured cable displacement
N = 2^nextpow2(NT);
Y = fft(Dy, N)/NT;
f = Fs/2*linspace(0, 1, N/2+1);     % Freq. pairs
symmetric about Nyquist freq.
Mag = 2*abs(Y(1:N/2+1));            % Magnitude multiply
by 2 to cancel the conjugate component

figure (11)
plot(f,Mag)                         % Plot FFT results
xlim([0.1,15])
xlabel('Frequency [Hz]', 'FontName','Arial','fontsize',
14);
ylabel('FFT Amplitude', 'FontName','Arial','fontsize',
14);
title('FFT of Measured Displacement','FontName','Arial
','fontsize',14);
set(gca,'FontName','Arial', 'fontsize', 14);
set(gcf,'Position',[800 200 800 600]);

%--------------- Cable tension calculation by the
Vibration Method
[f_n,Mag_n]=ginput(1);              % Select cable frequency
and order for tension calculation
close all
f_ind=input('Input mode order of the selected frequency : ');

% Cable geometric parameters: length(m), Diameter(m),
Area(m^2)
Params=[192.42*0.3048,0.128,0.011551];
Density=76.97/9.8*1000;            % Cable density (kg/m^3)
E_modulus=1.63e8;                  % Cable elastic modulus
(KN/m^2)
Density_PL=Density*Params(3); % Density per length
(units: Kg/m)
EI=E_modulus*pi*Params(2)^4/64;
```

```
phi_n=sqrt(EI/(Density_PL*(2*pi*f_n)^2*Params(1)^4));
% Introduced parameter phi_n

if f_ind==1        % Introduced parameter B for gamma_n
   B_ind=290;
else
   B_ind=0;
end
A_ind=-18.9+26.2*f_ind+15.1*f_ind^2;   % Introduced
parameter A for gamma_n
gamma_n=f_ind*pi+A_ind*phi_n+B_ind*phi_n^2;

Cable_force_KN=(4*pi^2*Density_PL*Params(1)^2*f_n^2/
gamma_n^2-EI*gamma_n^2/Params(1)^2)/1000   % Cable
force (units: KN)
```

7.3 Implementation in the Bronx-Whitestone Bridge Suspender Replacement Project

Suspenders are among the most critical members of a suspension bridge, and they suffer from corrosion-induced degradation during their service life. Practices in North America show that the average replacement of suspenders is roughly between 60 and 75 years. The Triborough Bridge and Tunnel Authority (TBTA) in New York City has a master plan for rehabilitating the Bronx-Whitestone Bridge, which includes investigating its suspender rope system for the purpose of assessing the condition and evaluating the strength of these suspender ropes. The work included an in-depth field inspection of selected suspenders and laboratory testing. In the field, inspectors looked for damage, corrosion, and other anomalies that might contribute to a reduction in their ultimate strength.

7.3.1 Bronx-Whitestone Bridge

The Bronx-Whitestone Bridge, shown in Figure 7.15, spans the East River connecting the boroughs of Queens and the Bronx in New York City. It is an important link in the regional transportation system and carries Interstate 678. The bridge was opened to traffic in 1939; the suspenders are original and therefore have been in service for nearly 80 years. The suspended span consists of a 2300 ft. main span and two 735 ft. side spans, all situated between land-based approaches. The towers are constructed of structural steel on concrete pedestals situated in water [59]. The concrete anchorages are both land-based.

Figure 7.15 Bronx-Whitestone Bridge.

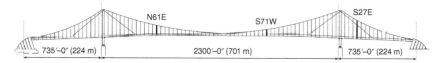

Figure 7.16 Suspender replacement locations.

Three suspender ropes were selected for removal and replacement based on the results of field inspection. The locations of these three suspender ropes are indicated in Figure 7.16. The two main suspension cables are identified as East (E) and West (W). Further identification is given by the side of the bridge measured from panel point 93 at mid-span. The Bronx side is the North (N) side, and the Queens side is the South side (S). Finally, the rope legs are designated as inward for the leg closest to the roadway and outward for legs closest to the river. The suspenders being replaced are identified as N61E, S71W, and S27E.

7.3.2   Estimating Suspender Tension

The three existing/original suspenders (i.e. N61E, S71W, and S27E) were successfully replaced between November 2016 and June 2017. Figure 7.17a shows the jacking apparatus that releases the tension in the temporary suspender ropes. As the jacks are released, the new suspender ropes are effectively tensioned (Figure 7.17b). Each suspender contains four legs (i.e. ropes).

The computer vision sensor is applied to measure the suspender ropes' vibrations and estimate their tension forces, in order to ensure that the tension forces of the new suspender ropes are within the desirable range. For the vision sensor system shown in Figure 7.18, the adopted video camera (Point Grey, FL3-U3-13Y3M-C) has a CMOS-type sensor with a maximum resolution of 1280×1024

(a) (b)

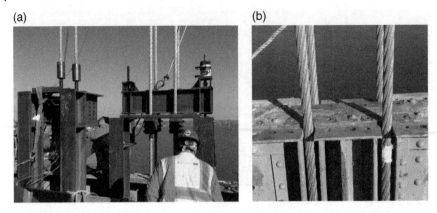

Figure 7.17 Field suspender replacement: (a) jacking apparatus; (b) new tensioned suspender ropes.

Figure 7.18 Vision sensor setup for measuring suspender tension. *Source:* Reproduced with permission of ASCE Library.

pixels and maximum rate of 150 fps, and the focal length of the optical lens varies in the range 16–160 mm with manual focus. Here, a sampling rate of 50 fps is adopted. The vibration of the four suspender legs (i.e. SL1, SL2, SL3, and SL4) can be simultaneously measured using one camera in real time, with the rope segments framed by red squares registered as templates for motion tracking.

Table 7.2 Cable geometric and material parameters.

Suspender	Effective length (m)	Diameter (m)	Elastic modulus (GPa)	Mass density (Kg/m)
N61E	21.95	0.051	137.9	10.63
S71W	9.45	0.051	137.9	10.63
S27E	34.44	0.051	137.9	10.63

Table 7.3 Measured suspender rope frequencies and tension.

	N61E					S71W	S27E
	SL1	SL2	SL3	SL4	Total 4 legs	Total 4 legs	Total 4 legs
First freq. (Hz)	3.40	3.34	3.28	3.28	NA	NA	NA
Tension (kips)	53	51	49	49	**202**	**204**	**256**

The geometric and material parameters for each leg of the three suspenders are tabulated in Table 7.2. The measured forces for the new suspender (four legs total) after replacement are listed in Table 7.3, and they agree well with the reference values provided by the design engineer. For simplicity, only the measured vibration time histories and their PSD amplitudes for each of the four legs of suspender N61E are plotted, as shown in Figure 7.19. The measured first natural frequency and rope force for each leg are listed in Table 7.3.

7.4 Summary

Cables are the most important components in cable-supported bridges and roof structures. Existing vibration methods for estimating cable tension are based primarily on measured acceleration responses. This practice is relatively expensive and time-consuming due to the required installation of contact-type sensors. The computer vision sensor successfully addressed such difficulties, as demonstrated in its field implementation in measuring the cable tension forces in two structures: the cable-supported roof structure of the Hard Rock Stadium in Florida and the suspension Bronx-Whitestone Bridge in New York City. The noncontact, remote measurement capability of the vision sensor eliminated the need to access the cables to install sensors, as required by conventional methods. The cable force was estimated based on the cable's natural frequency, identified from the measured displacements. There was no need to determine a scaling factor to convert

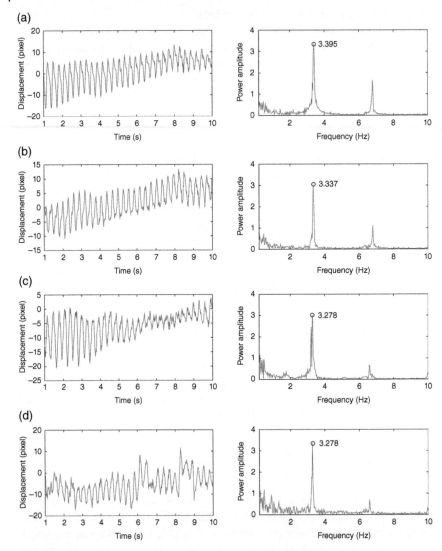

Figure 7.19 Measured vibration time histories and PSD amplitudes for suspender N61E: (a) suspender leg SL1; (b) suspender leg SL2; (c) suspender leg SL3; (d) suspender leg SL4.

pixel displacements to physical displacements, which further simplified the measurement procedure. The estimated cable forces based on the vision sensor measurements agreed well with those measured by jacking load cells, validating the accuracy of this computer vision sensor-based method. The successful implementation in these two engineering projects demonstrated that the computer vision sensor can be a highly cost-effective tool for temporary measurement and even long-term monitoring of cable forces in cable-supported structures.

8

Achievements, Challenges, and Opportunities

Rapid advances in computer vision science and technology have inspired the development of innovative computer vision-based solutions to challenging problems associated with civil engineering structures and infrastructure systems. In recent years, computer vision sensing has been drawing attention and gaining popularity in three major areas in the civil and structural engineering community: (i) vision-based sensors for structural displacement measurement and structural health monitoring (SHM) applications, which is the focus of this book; (ii) automated computer visual inspection and condition assessment, with or without using unmanned aerial vehicles (UAVs); and (iii) onsite construction tracking and safety monitoring. This concluding chapter summarizes the capabilities and open challenges of computer vision sensors based on the authors' work focusing on structural displacement response measurement and SHM. This chapter is also intended to share with readers the collective achievements of the research community in all of the three major areas by providing a state-of-the-art literature review. Challenges and opportunities are outlined to assist the structural engineering and computer science research communities in setting an agenda for future research.

8.1 Capabilities of Vision-Based Displacement Sensors: A Summary

This book introduces the fundamental principles of vision-based displacement sensor technology based on template-matching and target-tracking techniques. Extensive laboratory and field tests validate the measurement accuracy and robustness and demonstrate the unique advantages and significant potential of computer

Computer Vision for Structural Dynamics and Health Monitoring, First Edition.
Dongming Feng and Maria Q. Feng.
© 2021 John Wiley & Sons Ltd.
This Work is a co-publication between John Wiley & Sons Ltd and ASME Press.
Companion website: www.wiley.com/go/feng/structuralhealthmonitoring

vision sensors for advancing SHM research and practice. Some key issues of computer vision-based displacement measurements are summarized in this section.

8.1.1 Artificial vs. Natural Targets

To improve the robustness of target tracking and the accuracy of displacement measurements, high-contrast artificial targets are often attached to the structural surface [21, 60], such as roundel targets [61], concentric rings [27], crosses [62], LEDs [63], black-and-white blocks of random sizes [64], speckle patterns [65], etc. For example, in the literature [66], a planar target with four circles is attached to the structure to measure the vibration of a bridge; while in [62], cross-shaped targets are used, and the viewing system is equipped with an additional reference system, which decreases sensitivity to vibrations. Ring-shaped and random targets are used in [67], and multiple targets are simultaneously measured with a single camera, producing displacements at multiple points. To completely eliminate the need for accessing the structure, efforts have been made to track natural features on the structural surface without installing artificial targets. Recent studies have achieved accurate measurements by tracking natural structural surface features [68–70]. Robust template-matching algorithms, such as the orientation code matching (OCM) presented in this book, enable accurate measurement of bridge displacements from hundreds of meters away in low lighting conditions by tracking such natural surface features as rivets and edges, as reported by Fukuda et al. [32] and Feng and Feng [71], both of which are included in this book.

8.1.2 Single-Point vs. Multipoint Measurements

While zooming the camera lens to focus on a single point can maximize the resolution of the measured displacement, it is desired to use one camera to simultaneously measure structural displacements at multiple points across the structure. This requires zooming out the camera lens to increase the field of view, i.e. the area that is visible in the image, in order to include all the target points. As a result, the displacement resolution at each measurement point is reduced. Therefore, trade-offs between measurement resolution and the size of the field of view are necessary.

It is obviously not feasible to use one camera to measure all the points on a large structure with sufficient displacement resolution. A camera's field of view is limited. To monitor a large structure such as a long-span bridge, one possible solution is to use multiple synchronized cameras, with each camera targeting different sections of the structure. For example, in an effort to measure multipoint displacements of a large structure, Fukuda et al. [66] developed a time-synchronous measurement system using multiple computer and camera subsystems. The system embeds an algorithm that automatically and periodically performs synchronization through TCP/IP communication to maintain the time lag between the internal clocks of multiple computers in a range less than 5 ms. Lee et al. [72] introduced the synchronized multipoint vision-based system for real-time displacement measurements of high-rise buildings using a portioning approach, the feasibility of which was verified on a five-story steel frame tower. Ojio et al. [73] synchronized two cameras by triggering an interval timer through a relay module driven by solid-state relays. A pair of high-luminance LEDs were activated by the timer within sight of each camera and used as a synchronizing timing marker. Alternatively, one camera can be used to "scan" a large structure, i.e. to measure displacements segment by segment. For example, to extract frequencies and mode shapes of a large structure, Hoskere et al. [74] used a UAV to survey one portion of the structure at a time and obtain local mode shapes from each of the video records and then stitched them together using an ordinary least-squares fit.

8.1.3 Pixel vs. Subpixel Resolution

The computer vision sensor measures displacements with integer-pixel resolution, as the minimum unit in an image is one pixel. Although pixel-level resolution is adequate in many applications, higher resolution is often required to measure multipoint displacements with a wide field of view, as discussed earlier. In particular, pixel-level template matching may result in unacceptable measurement errors if the displacement to be measured is in the same order of magnitude as the scaling factor. To improve measurement resolution, subpixel registration techniques can be incorporated into the template-matching algorithms. Interpolation is the most commonly used subpixel approach, examples of which include intensity interpolation, correlation coefficient curve-fitting or interpolation, phase correlation interpolation, and geometric methods [36, 75, 76].

Subpixel registration can also be formulated as an optimization problem and solved through heuristic algorithms such as genetic algorithms, artificial neural network algorithms, and particle swarm optimization. There are also other subpixel techniques based on the Newton–Raphson method and gradient-based methods. For example, to further improve the accuracy of digital image correlation (DIC), Pan et al. [77] reviewed the various subpixel registration algorithms, including the coarse-fine search algorithm, peak-finding algorithm, iterative spatial domain cross-correlation algorithm, spatial-gradient-based algorithm, genetic algorithm, finite element method, and B-spline algorithm. Debella-Gilo and Kääb [78] evaluated two different approaches – intensity interpolation and correlation interpolation – to achieve subpixel precisions when measuring surface displacements on mass movements using normalized cross-correlation (NCC). Through a shaking table test, Feng et al. [28] demonstrated the improvement of measurement accuracy by applying an upsampling subpixel technique, which is presented in this book.

In practice, subpixel resolution is limited, as images may be contaminated with environmental noise and system noise arising from the electronics of the imaging digitizer. The subpixel accuracies reported in many studies vary within orders of magnitude from 0.5 to 0.01 pixels [36]. Ferrer et al. [79] demonstrated, through numerical analysis, a realistic limit for subpixel accuracy, and found that the maximum achievable resolution enhancement is related to the dynamic range of the image.

8.1.4   2D vs. 3D Measurements

The experiments presented in this book are mainly for 1D and 2D in-plane displacement measurements. Using a single camera for 2D in-plane measurements, however, the accuracy of the displacement measurements is sensitive to out-of-plane motion. Sutton et al. [80] found that the in-plane measurement error due to out-of-plane translation is proportional to $\Delta Z/Z$, where ΔZ is the out-of-plane translation displacement and Z is the distance from the object to the camera.

To minimize the effect of out-of-plane motion, a 3D stereovision system with a pair of synchronized cameras can be employed. Park et al. [25] proposed a motion capture system with multiple cameras to measure 3D structural displacements. The 2D coordinate data from each camera is used to calculate the 3D coordinates of the markers attached to structures with respect to the predetermined origin of the 3D space. Poozesh et al. [17] assessed the accuracy of the 3D stereovision system to measure full-field distributed strain and displacement over a large area of a scaled wind turbine blade. Pan et al. [77] pointed out that to measure the deformation of a macroscopic object with a curved surface, stereovision-based 3D measurement is more practical and effective because it can be used for 3D profile and deformation measurements and is insensitive to out-of-plane displacement.

Three-dimensional measurement based on stereovision systems is expected to attract more research and application interest. However, due to its convenience and efficiency, 2D measurement is sufficient for most civil engineering structural applications, such as measuring the vertical and transverse deformation of bridges and horizontal displacements of buildings or towers.

8.1.5   Real Time vs. Post Processing

Depending on the intended applications, it may be acceptable to extract structure displacement time histories from recorded video images via post-processing, as is done in most existing vision sensor systems reported in the literature. In this case, only a consumer-grade off-the-shelf video camera is needed to take and record videos. Post-processing also provides the flexibility to extract structural displacements at any measurement point from a single video recording.

However, for real-time continuous monitoring, the displacement must be measured by processing the video images in real time. Sometimes, only real-time displacement measurement data is saved in the computer, avoiding the time-consuming and memory-intensive task of saving video files. A real-time computer vision sensor system contains a video camera and a computing unit (such as a laptop computer) with an image-processing software package installed. The feasibility of real-time measurement depends on the number of measurement points, required measurement resolution, frame rate of the camera, sizes of the template and region of interest, and computer CPU speed. To achieve real-time measurements, the computer vision sensor algorithm and software must be made computationally efficient. Khuc and Catbas [68] mentioned that the challenges associated with the video storage requirement and processing time need to be addressed. Baqersad et al. [81] also pointed out that real-time capability is critical to make photogrammetry more practical for dynamic measurement. The feasibility of real-time measurement depends on the complexity of the adopted template-matching algorithm, the programming language, as well as the code efficiency for the developed software. Pan et al. [21] developed a real-time displacement-tracking system using subset-based DIC.

The authors of this book developed two real-time displacement measurement software packages based on the upsampled cross-correlation (UCC) and OCM template algorithms, respectively, by incorporating such techniques as confining a region of interest [28, 31]. The programming environment for the software is Visual Studio 2010 using the C++ language. However, in order to assist readers' self-learning, the authors have translated these computer codes into MATLAB and included it in this book.

8.2   Sources of Error in Vision-Based Displacement Sensors

Measurement errors, caused by various sources, cannot be completely eliminated for the computer vision sensor, as acknowledged by many studies, particularly those involving outdoor field applications. Measurement errors may arise from sources such as the calibration procedure, optical distortion effects, nonlinearity of the field view, system resolution, time synchronization among cameras, poor or changing illumination, and non-uniform air refraction, among others. D'Emilia et al. [60] evaluated the performance of vision-based vibration measurements by considering the effect of peculiar parameters, i.e. the type of target, vibration frequency and amplitude, exposure time, and image-acquisition frequency. Some studies [82, 83] have analyzed the sensitivity of displacements to image-acquisition noise related to digitization, read-out, black current, and photon noise, and demonstrated that the standard deviation of measurement errors is proportional to that of the image noise and inversely proportional to the subset size and to the average of the squared gray-level gradients. Haddadi [84] investigated, based on numerical and experimental tests of rigid-body motion, the error sources of the DIC technique in relation to lighting, optical lens (distortion), CCD sensor, out-of-plane displacement, speckle pattern, grid pitch, size of the subset, and correlation algorithm. Pan et al. [77] systematically reviewed the displacement measurement errors in 2D digital image correlations caused by speckle patterns, camera sensors not parallel to the object surface, out-of-plane displacement, image distortion, various noises, subset size, correlation criterion, interpolation scheme, shape function, etc. Ferrer et al. [79] performed a parametric study of measurement errors introduced by the vision-based method, in which influencing factors such as the distance to the target, image size, type of camera, and displacement amplitude were analyzed. Major sources of measurement errors related to the practical application of the computer vision sensor are discussed in more detail next.

8.2.1 Camera Motion

In field applications, the camera itself is often subjected to ambient vibrations due to wind and nearby traffic, causing displacement measurement errors, as pointed out in a number of studies. For example, during the field tests in [64], camera vibrations caused by passing-train-induced ground motion affected measurement accuracy. Measurement errors become more severe when a zoom lens is used, which magnifies not only the images but also the camera vibration. A lightweight, compact camera-tripod system is more prone to camera motion.

Yoneyama and Ueda [65] proposed a method for correcting the effect of camera movement in which the relationship between images before and after the camera movement is described by an equation of perspective transformation and the unknown coefficients are determined from un-deformed regions of the images. The effectiveness of the perspective transformation in correcting the camera movement is demonstrated in bridge-deflection measurements during vehicle passage. The authors of this book and others developed a more convenient method of canceling camera vibrations [71, 85, 86]. When measuring, for example, the mid-span vertical displacement of a bridge deck, a reference object (such as a building) in the background can be assumed stationary in the vertical direction. By subtracting the displacement of the reference object from the bridge displacement, the camera motion can be canceled [71]. Kim et al. [85] applied this correction method to remove deck-vibration-induced camera motion when measuring the hanger cable vibration of a bridge. However, the requirement for a stationary reference in the field of view may limit the application of this method in the field. Therefore, the effect of camera motion on displacement measurement accuracy is still an open problem.

8.2.2 Coordinate Conversion

The scaling factor determined by either of the two coordinate conversion methods described in this book can introduce measurement errors. For the camera calibration method in Chapter 2, the extrinsic and intrinsic matrices for camera calibration do not account for lens distortion: mostly radial distortion, plus slight tangential distortion. Lava et al. [87] and Pan et al. [88] investigated the impact of lens distortion on DIC. For accurate camera calibration, a general lens-distortion model should be considered. In the literature [89], a method for correcting lens distortion is proposed to improve the measurement accuracy of a 2D displacement vision sensor. The lens distortion is first evaluated from displacement distributions obtained in a rigid-body in-plane translation and rotation test, and the measured displacement is then corrected using a coefficient determined by a least squares method. Note that in field measurements of small structural displacements with a relatively long focal-length lens from a remote distance, errors caused by lens distortion may be negligible.

For the practical calibration method described in Chapter 2, there are also errors in the estimated scaling factor, arising from uncertainties in the camera tilt angle and camera distance measurements and focal length readings from the adjustable-focal-length lens. Using a fixed zoom lens can minimize errors in focal length reading. For the scaling factor SF_1 based on a known physical dimension on the target surface and the corresponding image dimension in pixels, errors in the range of ± 2 pixels could occur when manually selecting physical members in the image using a mouse. It is recommended to use the mean value from several repeated picking operations to average out some of the random errors. Furthermore, if a dot-extraction algorithm is implemented, the error can be reduced to $\pm 1/10$th of a pixel. The theoretical study in Chapter 2 investigates the effects of the optical axis tilt angle and lens focal length based on 1D in-plane translation. It is found that errors in scaling factors increase as the tilt angle increases, and the error is inversely related to focal length. Furthermore, the tests in Section 3.4 of Chapter 3 demonstrates that the estimated scaling factor utilizing known physical dimensions can yield satisfactory accuracy when the camera tilt angle is small. It is also suggested that for a non-perpendicular optical axis in a lens, scaling factors in the horizontal and vertical directions should be obtained separately. Moreover, when a series of targets are tracked simultaneously by a single camera to measure displacements at multiple points along the structure, due to projective distortion, different scaling factors should be determined for each measurement point using a known dimension closer to or encompassing that point.

8.2.3 Hardware Limitations

Measurement errors of computer vision sensors can also arise from hardware limitations such as the rolling shutter effect and temporal aliasing associated with low frame rates. Two types of image sensors widely used in digital cameras are CCD and CMOS sensors. Although CMOS image sensors show improvement over CCD, one major distinction between the two is the read-out modes. CCD cameras often use global shutter mode, which captures the entire image frame at the same instant. This is particularly beneficial when the image changes from frame to frame. In contrast, the majority of CMOS cameras in the consumer market use rolling shutter mode, in which each image frame is recorded by scanning the pixels row by row or column by column. Thus the rolling shutter method may produce distortions when recording fast-moving objects, the effect of which on measurement accuracy should be rectified. Note that not all CMOS sensors have rolling shutters. The video camera (Point Grey, FL3-U3-13Y3M-C) used in the various tests in this book has a CMOS-type sensor but a global shutter.

Temporal aliasing is another concern for vibration measurements. When the structural displacement contains frequency components greater than half the camera frame rate, the measured displacement will contain error information aliased from higher frequencies. Unlike conventional vibration sensor systems where the aliasing effect can be removed by applying an anti-aliasing filter, temporal aliasing cannot be mitigated with such filters in computer vision systems since the images are already aliased [69]. According to the Nyquist theorem, in order to avoid aliasing, the sample rate must be at least twice the highest frequency component in the measured signal. For field measurement of a structure whose natural frequencies are unknown, preliminary analysis can be conducted to estimate the natural frequencies. Note that for civil engineering structures, the dominant natural frequencies are usually less than 50 Hz, and thus a sampling rate of 100 Hz should be sufficient to avoid the aliasing problem.

8.2.4 Environmental Sources

It is well known that the accuracy of template matching is largely dependent on image quality, which is often difficult to guarantee in outdoor field environmental conditions. Illumination variation, partial target occlusion, partial shading, background disturbances, rain, snow, and haze present challenges. Shadows and low lighting conditions are the fundamental technical limitation of computer vision systems. For high-speed, high-magnification and/or night-time measurements, proper illumination of the target area is important. Measurement errors can also arise from the heat haze that occurs when the air is heated non-uniformly by high ambient temperatures during field testing. The non-uniformly heated air causes variations in its optical refraction index, resulting in image distortion. Measurement errors caused by heat haze increase as the measurement distance increases. Research has been conducted to study measurement errors from these environmental sources. For example, Ye et al. [19] conducted a series of shaking table experiments to examine environmental influence factors affecting the accuracy and stability of the vision-based system. It is demonstrated that measurement results are adversely affected by illumination and vapor. Ribeiro et al. [90] have shown that the measurement precision of a video system can be affected by distortion of the field of view, due to the flow of heat waves generated by IR incandescent lighting and that, therefore, the operating time of the lamps should be limited. Lee et al. [91] pointed out that external factors such as precipitation, fog, variation of natural light, and wind action may influence the performance of a vision system. Anantrasirichai et al. [92] proposed a novel method for mitigating the effects of atmospheric distortion using complex wavelet-based fusion. As demonstrated in this book, the gradient-based OCM template-matching algorithm can mitigate many of the environmental effects and thus is suited for robust and accurate monitoring of outdoor engineering structures. The displacement measurement error caused by haze-induced image distortion is still an open challenge.

8.3 Vision-Based Displacement Sensors for Structural Health Monitoring

With its unique advantages over conventional sensors, the computer vision sensor is expected to advance SHM technology toward practical, widespread implementations. Many studies have been conducted along this line, as reported in the literature to be reviewed in this section. This book has also presented a number of examples of applications of the computer vision measurement of structural responses for modal analysis, updating structural models, and detecting damage.

8.3.1 Dynamic Displacement Measurement

Many recent studies of computer vision sensors have experimentally demonstrated that high accuracy can be achieved for both single-point and multipoint structural displacement measurements by tracking either high-contrast predesigned artificial targets or existing natural features on the structural surface [27, 63, 67, 68, 90, 91, 93–99]. D'Emilia et al. [60] introduced the concept of synchronizing cameras with laser sensors and accelerometers. Recently, efforts have also been made to investigate the feasibility of displacement measurements utilizing advanced smartphone technology, including an embedded high-resolution/speed video camera, powerful processor and memory, and open source computer vision libraries. For example, Min et al. [100] developed a smartphone software application for real-time displacement measurement, and shaking table tests were conducted to evaluate its accuracy. Based on vibration testing of a small-scale multistory laboratory model, Ekin et al. [101] demonstrated the dual usage of ubiquitously available smartphones to measure both structural deflections/displacements and accelerations with their embedded cameras and accelerometers. Warren et al. [102] experimentally compared vision-based, laser, and accelerometer measurements for analyzing structural dynamics and concluded that vision-based photogrammetry techniques favor high-amplitude, low-frequency vibration, which is typically difficult to measure with accelerometers and laser vibrometers. As presented in Chapters 3 and 4 of this book, the authors carried out shaking table and impact tests to demonstrate the accuracy and robustness of the computer vision sensor in measuring structural dynamic displacements containing a wide range of frequency components.

For civil engineering structural applications, field evaluation of computer vision sensors in outdoor environments is of particular importance. For example, field tests individually conducted by Feng et al. [28] and Shariati and Schumacher [103] on a pedestrian bridge located on the Princeton University campus cross-validated the frequency-domain characteristics of the bridge identified from measurements by vision sensors. Pan et al. [21] demonstrated the practicality of a video deflectometer for real-time deflection measurement of a railway bridge. Ribeiro et al. [90] compared train-induced deck displacements of a railway bridge measured by a vision sensor and a linear variable differential transformer (LVDT), demonstrating measurement errors less than 0.25 mm when the camera was placed 25 m away. Busca et al. [67] proposed a vision-based technique to measure both the static and dynamic displacement responses of a railway bridge and found that the measurement reliability is strongly affected by the structure's texture contrast when the measurement was conducted without using artificial targets and instead relying only on the natural texture of the bridge. This study echoed that in order to measure a large portion of a bridge with a single camera, a compromise between the field of view and measurement resolution is necessary.

As presented in the previous chapters, the authors carried out field tests on a number of short-span bridges, including a stiff highway bridge under vehicle loads and two railway bridges under trainloads at various speeds. The tests validated the accuracy of the computer vision sensor system in measuring low-amplitude (as low as 1 mm) bridge response displacement and the robustness of tracking natural targets such as rivets even at night with dim illumination.

Studies have also been carried out on remote measurements of long-span bridges and other large structures. Stephen et al. [61] employed a visual tracking system in measuring deck displacements at the center of the 1410 m span of the Humber Bridge in the UK. Brownjohn et al. [27] evaluated the performance of a commercial camera system for tracking mid-span displacement of the Humber Bridge by using both pre-attached artificial target panels and natural targets on the bridge, with the camera placed 710 m from the targets. The measurements agree well with data from reference GPS. Ye et al. [104] demonstrated the robustness of their vision sensor system through field measurement of the mid-span vertical displacement of Tsing Ma Bridge, and good agreement was observed between measurement results from the vision-based system and GPS. In addition, the vision sensor system was used to measure the vertical mid-span displacement influence lines of the Stonecutters Bridge in Hong Kong under different loading scenarios. By tracking six actively illuminated LED targets mounted on the bridge, the video deflectometer developed by Pan et al. [21] was applied for remote and multipoint displacement measurements of the Wuhan Yangtze River Bridge in China during its routine safety evaluation tests. In [68], the measurement accuracy of the vision sensor was validated by comparing the modal parameters of a football stadium identified from measurements by the vision sensor and reference LVDTs and accelerometers, respectively, in changing ambient light and at various camera distances. Wahbeh et al. [63] developed a video camera system with targets consisting of black steel sheets on which two high-resolution red lights (LED) were mounted to measure the displacement of the 1847 m-long Vincent Thomas Bridge located in Sam Pedro, California. As presented in this book, the authors conducted vision sensor-based displacement measurement of the same bridge by tracking the natural targets (rivets) on the bridge from 300 m away, demonstrating the robustness of the OCM algorithm in challenging field conditions including long camera distance, low-contrast natural targets, and changing natural illumination. The author also applied the computer vision sensor to the 2089 m-long Manhattan Bridge in New York City, confirming the sensor's capabilities in remote, real-time, multipoint measurements [71] and successfully addressing the camera vibration issue with the practical cancelation method based on simultaneous measurement of a stationary reference target.

8.3.2 Modal Property Identification

The computer vision sensor system is a cost-effective tool for experimental modal analysis to identify modal parameters, including frequencies and mode shapes, and study dynamic properties of structural systems in the frequency domain. In fact, structural health monitoring methods are often based on vibration measurements and analysis of changes in dynamic structural properties. Conventional vibration sensors, such as accelerometers, GPS, and laser vibrometers, are point sensors, most of which must be installed on the structure. Obviously, the spatial resolution of the obtained mode shapes is limited by the number of deployed sensors; and the locations of the sensors, once installed, cannot be easily altered.

Ferrer et al. [79] developed a method for simultaneous multipoint measurements of vibration frequencies through the analysis of a high-speed video sequence. Wang et al. [105] carried out full-field vibration measurements on the 3D surface of a car hood with a 3D DIC system with random excitation, from which modal parameters of the hood were successfully identified from the frequency response functions. Yoon et al. [69] carried out laboratory vibration experiments on a scaled frame structure to show the usefulness of the computer vision sensor for modal analysis. Poozesh et al. [17] extracted mode shapes and natural frequencies of a small-scale wind turbine blade by using two synchronized stereovision systems in conjunction with output-only system identification. The modal properties extracted from stitched measurements by camera pairs were shown to be accurate when compared to those from a validated finite element model, validating the stitching approach for a multi-camera system for monitoring a large structure. To obtain mode shapes of full-scale civil infrastructure, Hoskere et al. [74] used a UAV to take a video of the structure one segment at a time and obtain the global mode shapes of the structure by stitching together the local mode shapes from each recorded segment using an ordinary least squares fit.

As presented in Chapter 4, the authors conducted laboratory experimental modal analysis on a scaled simply supported beam and a three-story frame structure, demonstrating that the natural frequencies and mode shapes of these structures identified from simultaneously measured multipoint displacements by a single camera agreed well with those from conventional accelerometers installed on the structures. Furthermore, high-spatial-resolution mode shapes can be achieved by a single measurement with a single camera, which is a unique advantage of the computer vision sensor that cannot be matched by conventional sensors. Moreover, if only structural frequency and mode shape information is required from a dynamic test, coordinate transformation is unnecessary. In other words, there is no need to determine the scaling factor to transform pixel coordinate vibrations into physical coordinate ones, which makes the vision-based measurement procedure even more efficient.

8.3.3 Model Updating and Damage Detection

An analytical model of a structure can be developed based on its design information; and the parameters of the model, such as stiffness, can be updated based on vibration measurements and modal analysis of the structure, by minimizing the differences between the computed and measured modal properties, including the natural frequencies and mode shapes [106]. The updated model can then be used for structural damage detection and other purposes [40].

From the displacement of a cantilever beam measured by the phase-based optical flow algorithm, Cha et al. [107] utilized the unscented Kalman filter to detect structural damage by identifying structural properties such as stiffness and damping coefficients with an assumption of known structural mass. Oh et al. [108] updated models using a multi-objective optimization algorithm based on displacement responses measured by a camera-based motion capture system. Through a free vibration test of a three-story shear frame model, the model-updating method was validated by comparing the dynamic properties identified from the updated model and the response measurement. Wang et al. [109] demonstrated that the region-based Zernike moment descriptor is a robust image-processing technique for recognizing mode shapes and updating finite element models of simple plate structures. Furthermore, the nonlinear elastoplastic material properties were updated using the descriptor derived from full-field strain measurements [106]. Through laboratory experiments on steel cantilever beams, Song et al. [110] demonstrated a subpixel virtual visual sensor for acquiring modal shapes and frequencies of the structures and their use in a wavelet-based structural damage-detection algorithm. Dworakowski et al. [20] obtained the deflection curves of small-scale beams by means of DIC and evaluated two deflection shape-based algorithms for detecting damage in the beams.

In Chapter 4 of this book, the authors experimentally demonstrated that structural damage in a simple beam can be directly identified and located from a high-spatial-resolution mode-shape curvature index, and the computer vision sensor has the unique capability to construct such a smooth index. This book has also provided an example of identifying/updating inter-story stiffness values of a frame structure through frequency-domain optimization based on white noise-induced structural vibrations measured by the computer vision sensor. As presented in Chapter 5, for railway bridges subjected to trainloads that cannot be treated as white noise, the authors proposed a time-domain optimization method for identifying/updating structural properties and validated this method for a railway bridge based on displacements measured by the computer vision sensor and prior knowledge about the trainloads. Sensitivity studies showed that train-induced displacement responses are better suited than acceleration responses for identifying bridge stiffness.

In Chapter 6, the authors presented an iterative ordinary least squares solution method for simultaneously identifying structural properties and excitation force input based on measuring structural response output. Using this method, the stiffness of a beam structure as well as external impact excitation forces were successfully and accurately identified from the multipoint beam displacement responses measured by one camera.

8.3.4 Cable Force Estimation

A cable system is the most important component in cable-supported bridges and roof structures, and as such, the cables' tension needs to be monitored. A commonly used method is to estimate cable tension based on the cable's vibration frequency measured by accelerometers attached to the cable. Such a practice is relatively expensive and time-consuming due to the required installation of the sensor system. A few attempts have been made to apply vision-based sensors to estimate cable force. For example, Ji and Chang [111] and Kim et al. [85], respectively, employed the optical flow and normalized cross-correlation temptation methods for cable vibration measurements and force estimates. As described in Chapter 7, the authors have successfully implemented the computer vision sensor in two engineering projects to measure the force of the cables in a cable-supported roof structure of the Hard Rock Stadium in Florida and the Bronx-Whitestone Bridge in New York City. The results were validated by reference readings from load cells. To identify frequencies, the measured displacement in pixels does not need to be converted to physical units, making the low-cost, noncontact computer vision method even more attractive to implement for estimating cable tension and monitoring long-term health [2].

8.4 Other Civil and Structural Engineering Applications

8.4.1 Automated Machine Visual Inspection

Structural health monitoring methods based on structural vibration measurements, modal analysis, and model updating, as discussed previously, can identify and locate structural damage that results in losses of structural stiffness or changes in mode shape curvatures. By implementing continuous monitoring, any "hot spots" on the structure can be identified in a timely fashion, which can guide a targeted and condition-based (rather than the current condition-based) inspection of the structure. Computer vision can also be applied for such inspections to detect and assess damage to members of structures. Currently, civil engineering structures and infrastructure systems such as buildings, bridges, dams, tanks, and pipes are predominantly inspected by means of human visual inspection. As shown in an example in Figure 8.1a, such inspections are labor-intensive, time-consuming, expensive, subjective, and prone to human error. On the other hand, computer vision, combining artificial intelligence and UAVs, as shown in Figure 8.1b, offers significant potential for automated, objective, quantitative, low-cost machine visual inspection.

To date, much of the damage detection done using computer vision has focused on visible surface damage such as cracks, spalling, defective joints, corrosion, loosened bolts, potholes, etc., as shown in Figure 8.2. Koch et al. [112] reviewed the current state of practice of computer vision-based damage detection for civil infrastructure, in particular for reinforced concrete bridges, precast concrete tunnels,

(a)

(b)

Figure 8.1 Bridge inspection: (a) conventional visual inspection; (b) UAV inspection.

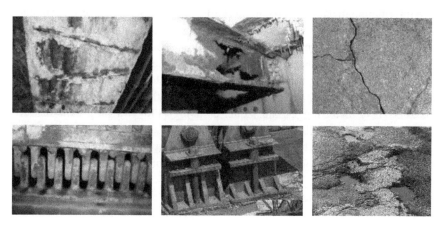

Figure 8.2 Examples of visible damage.

underground concrete pipes, and asphalt pavement. Mohan and Poobal [57] provided a detailed survey of vision-based crack-detection methods and a summary of various challenges in this area. Most recently, the use of camera-equipped UAVs for visual inspection and monitoring has grown exponentially. Providing excellent temporal and spatial resolution, this method is suitable for monitoring large structures and hard-to-reach areas [113]. The rapid technological advances in sensing, navigation, digital camera, computing, batteries, and aeronautics have helped make UAVs more affordable, reliable, and easy to operate [114]. Comprehensive literature reviews on computer vision applications in relation to UAVs, including achievements and future trends, can be found in [113–115].

In general, state-of-the-art computer vision systems have found some success in automated detection and quantification of damage, such as concrete cracks and spalling; cracks, holes, and joint damage in concrete pipes; and cracks, potholes, and patches in pavement. For inspection of long-span or large-scale structures such as bridges and tunnels, computer vision visualization techniques such as image-stitching have improved the spatial resolution of images, contributing to successful damage detection [112].

8.4.2 Onsite Construction Tracking and Safety Monitoring

Continuous monitoring and early warnings are essential to ensure construction safety, due to the inherently hazardous nature of construction sites. The common practice is to use safety personnel to conduct visual inspections and evaluate the risk involved in ongoing work and existing site conditions. However, such human visual observation methods are costly, and it is difficult to aggregate comprehensive information in a timely fashion. Recently, computer vision has been explored for automated and continuous monitoring at construction sites. Providing a rich set of information from images or videos, computer vision can facilitate the understanding of complex construction tasks rapidly, accurately, and comprehensively [116]. In particular, computer vision has been applied to various areas in construction such as monitoring ergonomic risks related to the posture of construction workers [117], detecting unsafe ladder-climbing [118], tracking construction workers to monitor project performance [119], automated detection of off-highway dump trucks [120], image-based automated safety assessment of earthmoving and surface mining activities [121], acquisition of 3D spatial coordinates of project entities [122], etc.

Despite the progress made in recent years, many technical and practical issues remain when applying diverse computer vision techniques to construction tracking and safety monitoring. For example, the reliability of the safety assessment results from the computer vision system is hindered by the lack of task-specific, quantitative metrics for unsafe conditions and actions. Due to the dynamic nature of construction sites, computer vision systems tend to suffer from such technical obstacles as object occlusions, improper camera positions, and lack of comprehensive image datasets with varied viewpoints for accurate assessment. In addition, workers' privacy issues must be addressed in order to implement computer vision for continuous monitoring at construction sites [116].

8.4.3 Vehicle Load Estimation

When using vehicle-induced bridge vibrations to identify the structural parameters of bridges, due to the difficulty of measuring vehicle loads, the input is often treated as spatially uncorrelated white noise so that output-only frequency-domain identification methods can be employed. Finding that such an assumption is incorrect for most highway bridges that are subjected to one-way traffic,

Chen et al. [44, 123] and Tan et al. [45] used video images of passing vehicles to extract information about the vehicle types, arrival times, and speeds, based on which they developed a stochastic model of vehicle loading excitations and proposed an output-only gray-box modal identification technique for bridge structures. They further incorporated this vehicle loading model into a Bayesian framework and demonstrated its effectiveness for updating a model of a highway bridge in California.

Other researchers have also attempted to apply computer vision to classify traveling vehicles and estimate their weights and positions. For example, Gandhi et al. [124] presented multisensory testbeds for collection, synchronization, archival, and analysis of multimodal data for monitoring the health of highway transportation infrastructures, where computer-vision algorithms are used to detect vehicles and extract their properties. Catbas et al. [125] presented a methodology for rating bridge loads by integrating computer images of passing vehicles with strain gauge measurements of the bridges. Ojio et al. [73] proposed a contactless vehicle weight-in-motion system by using one camera to track bridge deflections during the vehicle's passage and a second camera to determine the vehicle's position and axle spacing. The axle weights can be found by minimizing the differences between the measured and computed deflection responses if a reliable bridge analytical model is available. Recently, Feng et al [126] proposed a computer vision-based nonconatact vehicle weigh-in-motion method using the tire images taken by a roadside camera. The force exerted by a tire on the roadway surface is the product of the tire-roadway contact area and the inflation pressure. Computer vision is applied to measure the tire deformation/contact area and to find the manufacturer-recommended inflation pressure from recognizing the tire surface markings. This computer vision method enables, for the first time, a noncontact low-cost measurement of the weight and location of heavy vehicles on a bridge, which, combined with simultaneous measurement of bridge response displacement also by computer vision, opens up a new avenue of bridge structural health monitoring for more accurate diagnostics and prognostics of structural conditions.

8.4.4 Other Applications

Computer vision sensors have been applied to track movements on the surface of the earth such as glacier flow, rock-glacier creep, and landslides. For example, Debella-Gilo and Kääb [78] evaluated the accuracy of pixel and subpixel image-processing algorithms to measure mass surface movements. Computer vision has also been used for real-time monitoring of water levels without disturbing the water flow. For example, to characterize the dynamics (including frequencies and damping ratios) of tuned liquid column dampers, a vision-based sensing system was developed to measure the water depth time history during seismic shaking table tests [127].

In the field of experimental mechanics, including mechanical testing of materials and structural stress analysis, the digital image correlation technique has been commonly used as a practical and effective tool; it directly provides full-field displacements and strains by comparing digital images of the test object surface acquired before and after deformation. Experimental mechanics applications usually involve testing specimens, and computer vision measurements are made in well-controlled laboratory environments. To achieve reliable and accurate measurement results, artificial speckle or texture patterns are often applied on the specimen surface for easy tracking. Pan et al. [77] systematically reviewed 2D digital image-correlation methodologies for displacement field measurement and strain field estimation and provided detailed analyses of measurement accuracy considering the influences of both experimental conditions and algorithm details. Based on the measured strain fields, various material mechanical parameters, including Young's modulus, Poisson's ratio, stress intensity factors, residual stress, and thermal expansion coefficients can be further identified.

8.5 Future Research Directions

Computer vision is generally recognized by both academia and industry as the next-generation smart sensing technology for health monitoring and condition assessment of structures and civil infrastructure systems. Despite the rapid progress of computer vision made in recent years, its applications for solving challenging civil and structural engineering problems are still at an early stage. Various technical and practical issues must be addressed in order to successfully employ computer vision technology for inspection and monitoring of large-scale structures with complex geometries in field environments. Future work may be aimed at addressing the following open research challenges and unsolved problems.

- Outdoor weather conditions such as insufficient lighting, shade, rain, snow, and heat haze cause image distortion and degradation, resulting in errors in displacement measurement, damage detection, and condition assessment. Ambient vibration of the camera also affects measurement accuracy. These various error sources pose particular challenges for computer vision sensors to accurately measure small-amplitude displacements and low-contrast surface damage/defects, especially when the camera is placed far from the targets. Heat haze remains an unsolved problem for outdoor remote measurement of structural dynamic response time histories. The capabilities of computer vision are limited when inspecting underground structures, such as tunnels and pipes, due to poor illumination, irregularly patterned backgrounds, and lack of contrast.

- While the capability of computer vision has been demonstrated in detecting structural surface damage and extracting properties such as the length, width, and orientation of a crack, it is equally important to determine the exact location of the damage. Therefore, it is imperative to map images to the global coordinate system, which is a challenging task.

- The computer vision sensor has demonstrated its potential for continuous monitoring of structural dynamics and health conditions as well as for autonomous machine visual inspection, but a significant gap exists between monitoring/inspection results and decision-making. How to transform the computer vision sensor data into actionable information for intelligent asset management remains a wide-open challenge.

Judging from the enormous enthusiasm from both the academic research and engineering practice communities, the authors are optimistic that next-generation computer vision, empowered by artificial intelligence and assisted by advanced robotics (such as UAVs), will be rapidly developed to facilitate widespread deployment of the SHM technology. In particular, the integration of computer vision-based continuous monitoring of overall structural dynamics and integrity with local autonomous machine visual inspection will enable timely, reliable damage detection and quantitative condition assessment of large structures and civil infrastructure systems. This will inspire a new paradigm of intelligent infrastructure maintenance and management to ensure structural functionality, safety, resiliency, and sustainability.

Appendix

Fundamentals of Digital Image Processing Using MATLAB

A.1 Digital Image Representation

There are primarily three types of images within the normal human perceptual range: binary, grayscale, and true color (Figure A.1). A *binary image* is the simplest type of image and containing only two values: 0 and 1, interpreted as black and white. A binary image is referred to as a *1-bit image* because it takes only one binary digit to represent each pixel. A *grayscale image* is also referred to as a *monochrome* or *intensity* image and consists of only shades of gray. A gray color is one in which the red, green, and blue components have equal intensity in RGB space, and thus it is only necessary to specify a single intensity value for each pixel. Often, the grayscale intensity is stored as an 8-bit integer giving 256 shades of gray from black to white (a pixel value of 0 corresponds to black, and a pixel value of 255 means white; intermediate values indicate varying shades of gray). Grayscale is particularly well suited to intensity images, which express only the intensity of the signal as a single value at each pixel in the region. A *true-color image*, also known as an *RGB image*, is an image in which each pixel is specified by three values – one each for the red, blue, and green components. The color of each pixel is determined by the combination of the red, green, and blue intensities stored in each color plane at the pixel's location. Graphics file formats store true-color images as 24-bit images where the red, green, and blue components are 8 bits each. This yields a potential 16 million colors.

A digital image is treated as a grid of discrete elements possessing both spatial (location) and intensity (color) information and ordered from top to bottom and left to right. An image containing one or more color bands that define the intensity or color can be represented by a two-dimensional (2D) integer array, or a

Computer Vision for Structural Dynamics and Health Monitoring, First Edition.
Dongming Feng and Maria Q. Feng.
© 2021 John Wiley & Sons Ltd.
This Work is a co-publication between John Wiley & Sons Ltd and ASME Press.
Companion website: www.wiley.com/go/feng/structuralhealthmonitoring

(a)   (b)

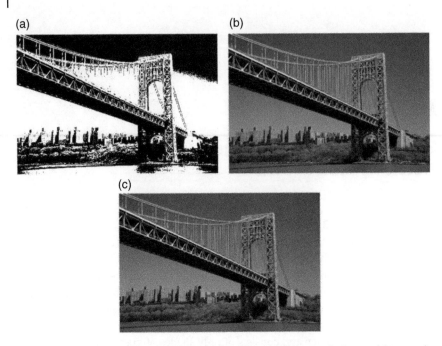

(c)

Figure A.1 Examples of image types: (a) binary image; (b) grayscale image; (c) true-color image.

Figure A.2 The 2D Cartesian coordinates of an M × N grayscale image.

series of 2D arrays, one for each color band. Each element of the array is called a *pixel*, derived from the term *picture element*.

In practice, grayscale images are sufficient for many computer vision–related tasks, and there is no need to use more complicated and harder-to-process color images. As illustrated in Figure A.2, the 2D discrete, grayscale image I(m, n) represents the response of the image sensor at a series of fixed positions ($m = 1,2, ...,$ M; $n = 1,2, ..., $ N) in 2D Cartesian coordinates. The indices m and n, respectively, designate the rows and columns of the image. Following the MATLAB convention, I(m, n) denotes the response of the pixel located at the mth row and nth column starting from a top-left image origin.

The following scripts introduce basic functions for reading, querying, converting, writing, and displaying different image types using MATLAB.

MATLAB Code – Basic Image Representation

```
%------------- Basic image manipulation
imfinfo('Bridge_RGB.jpg')              % query image
information such as FileSize, Format, ColorType, etc.

Bridge_RGB = imread('Bridge_RGB.jpg'); % read in the
jpg format image
imshow(Bridge_RGB);                    % display color image

Bridge_RGB(25,50, :)                   % print RGB pixel
value at location (25,50)
Bridge_RGB(25,50, 1)                   % print only the red
value at (25,50)

Bridge_red   = Bridge_RGB(:,:,1);   % extract red
channel (1st channel)
Bridge_green = Bridge_RGB(:,:,2);   % extract green
channel (2nd channel)
Bridge_blue = Bridge_RGB(:,:,3);    % extract blue
channel (3rd channel)

subplot(2,2,1); imshow(Bridge_RGB); axis image;
% display all in 2x2 plot
subplot(2,2,2); imshow(Bridge_red); title('Red');
subplot(2,2,3); imshow(Bridge_green); title('Green');
subplot(2,2,4); imshow(Bridge_blue); title('Blue');
```

```
Bridge_gray=rgb2gray(Bridge_RGB);    % convert truecolor
RGB image to a grayscale image
imshow(Bridge_gray);                 % display grayscale
image
Bridge_gray(25,50)                   % print single
pixel value at location (25,50)

subplot(1,2,1);
imshow(Bridge_RGB); axis image;      % display both RGB
and grayscale images side by side
subplot(1,2,2);
imshow(Bridge_gray);

imwrite(Bridge_gray,'Bridge_gray.jpg','jpg');
  % Write the grayscale image to disk as a JPEG image
imfinfo('Bridge_gray.jpg')
  % query the resulting grayscale image
```

A.2 Noise Removal

Digital images are prone to various types of noise. *Noise* is the result of errors in the image-acquisition process that result in pixel values that do not reflect the true intensities of the real scene. There are several ways that noise can be introduced into an image, depending on how the image is created. For example:

- If the image is acquired directly in a digital format, the mechanism for gathering the data (such as a CCD detector) can introduce noise.
- Electronic transmission of image data can introduce noise.
- If the image is scanned from a photograph made on film, the film grain is a source of noise. Noise can also be the result of damage to the film or can be introduced by the scanner.

To simulate the effects of some of the problems listed here, the MATLAB Image Processing toolbox provides the *imnoise* function, which can be used to add various types of noise to an image. The following scripts introduce basic functions for adding image noise and removing it using MATLAB. The results are illustrated in Figure A.3.

Original grayscale image

Grayscale image with Gaussian noise

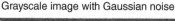

Noise removal with mean filtering

Noise removal with Gaussian filtering

Figure A.3　Noise-removal example.

MATLAB Code – Noise Removal

```
%------------- Filtering for image noise removal
Bridge_gray = imread('Bridge_gray.jpg');    % read in
the grayscale image
Bridge_gray_gn=imnoise(Bridge_gray,'gaussian',0.02);
  % add Gaussian noise with 0.02 variance

k=ones(3,3)/9;                 % define a 3x3 mean filter
Bridge_gray_gn_m=imfilter(Bridge_gray_gn,k);
  % apply mean filter to noisy image

k=fspecial('gaussian',[5 5],2);  % define a 5x5
Gaussian filter kernel
Bridge_gray_gn_g=imfilter(Bridge_gray_gn,k);
  % apply Gaussian filter to noisy image

figure
subplot(2,2,1); imshow(Bridge_gray); title('Original
grayscale image')
```

```
subplot(2,2,2); imshow(Bridge_gray_gn);
title('Grayscale image with Gaussian noises')
subplot(2,2,3); imshow(Bridge_gray_gn_m); title('Noise
removal by the mean filtering')
subplot(2,2,4); imshow(Bridge_gray_gn_g); title('Noise
removal by the Gaussian filtering')
```

A.3 Edge Detection

In a digital image, an *edge* is a curve that follows a path of rapid change in image intensity. Edges are often associated with the boundaries of objects in a scene. Edge detection is a fundamental tool in image processing and computer vision, particularly in the areas of feature detection and feature extraction.

In MATLAB, the function *edge* provides several derivative estimators to identify the edges in an image. Figure A.4 illustrates the detection results using different operators that *edge* provides. Among them, the most powerful edge-detection method is *Canny*. The *Canny* method differs from the other edge-detection

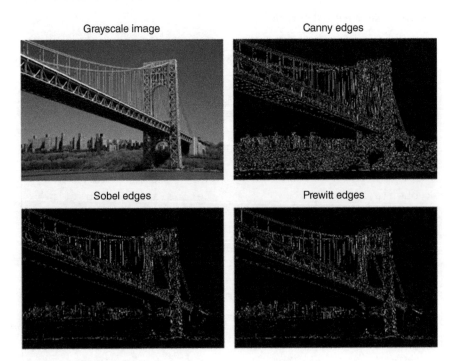

Figure A.4 Edge-detection example.

methods in that it uses two different thresholds (to detect strong and weak edges) and includes weak edges in the output only if they are connected to strong edges. This method is therefore less likely than the others to be affected by noise, and more likely to detect true weak edges.

MATLAB Code – Edge Detection

```
%------------- Image edge detection
Bridge_gray = imread('Bridge_gray.jpg');    % read in
the grayscale image
Bridge_Ec = edge(Bridge_gray,'canny');      % roberts
                                              edges
Bridge_Es = edge(Bridge_gray,'sobel');      % prewitt
                                              edges
Bridge_Ep = edge(Bridge_gray,'prewitt');    % sobel
                                              edges
figure
subplot(2,2,1); imshow(Bridge_gray); title('Grayscale
image')
subplot(2,2,2); imshow(Bridge_Ec); title('canny edges')
subplot(2,2,3); imshow(Bridge_Es); title('Sobel edges')
subplot(2,2,4); imshow(Bridge_Ep); title('Prewitt edges')
```

A.4 Discrete Fourier Transform

The Fourier transform is an important image-processing tool that is used to decompose an image into its sine and cosine components. In the Fourier domain image, each point represents a particular frequency contained in the spatial-domain image. The Fourier transform is used in a wide range of applications such as image analysis, image filtering, image reconstruction, and image compression.

The discrete Fourier transform (DFT) is a sampled Fourier transform and therefore does not contain all frequencies forming an image; rather, it includes only a set of samples large enough to fully describe the spatial-domain image. The number of frequencies corresponds to the number of pixels in the spatial domain, i.e. the images in the spatial and Fourier domain are the same size. For an image of size $M \times N$, the 2D DFT can be given by the expression

$$F(u,v) = \frac{1}{\sqrt{MN}} \sum_{x=0}^{M-1} \sum_{y=0}^{N-1} f(x,y) \exp\left[-2\pi i \left(\frac{ux}{M} + \frac{vy}{N}\right)\right] \tag{A.1}$$

where array $f(x, y)$ represents the spatial-domain image, and the exponential term is the base function corresponding to each point $F(u, v)$ in the frequency domain. The base functions are sine and cosine waves with increasing frequencies: i.e. $F(0, 0)$ is the DC-component of the image, which represents the average brightness. In most implementations, the frequency image is shifted in such a way that the image mean $F(0, 0)$ is displayed in the center of the image.

In a similar way, the frequency-domain image can be converted to the spatial domain. The inverse DFT, which has the same sample values as the original input sequence, is defined as:

$$f\left(x, y\right) = \frac{1}{\sqrt{MN}} \sum_{u=0}^{M-1} \sum_{v=0}^{N-1} F\left(u, v\right) \exp\left[+2\pi i\left(\frac{ux}{M} + \frac{vy}{N}\right)\right] \tag{A.2}$$

The Fourier transform produces a complex number–valued output image, which can be displayed with two images: either with the *real* and *imaginary* part or with *magnitude* and *phase*. In image processing, often only the magnitude of the Fourier transform is displayed, as it contains most of the information of the geometric structure of the spatial-domain image. However, if we want to re-transform the Fourier image into the correct spatial domain after some processing in

Figure A.5 Discrete Fourier transform of a grayscale image.

the frequency domain, we must make sure to preserve both the magnitude and phase of the Fourier image. Figure A.5 illustrates the real and imaginary parts of the Fourier image of the previous grayscale bridge image. Often, the magnitude plot of Fourier transforms displays a white dot in the middle surrounded by black because the DC component is much larger than other values. Here, the transform values are stretched by displaying a log of magnitudes.

MATLAB Code – Discrete Fourier Transform of a Grayscale Image

```
%------------- Perform 2D fast Fourier transform
Bridge_gray = imread('Bridge_gray.jpg');   % read in
the grayscale image
fft_Bridge = fft2(double(Bridge_gray));    % returns 2D
DFT
fft_Bridge = fftshift(fft_Bridge);         % shift
zero-frequency component to center of spectrum
figure(1)
subplot(2, 2, 1);imshow(Bridge_gray); title('Grayscale
image')
subplot(2, 2, 2);imshow(real(fft_Bridge));title('Real
Part of Spectrum')
subplot(2, 2, 3);imshow(imag(fft_
Bridge));title('Imaginary Part of Spectrum')
subplot(2, 2, 4);imshow(log(abs(fft_
Bridge)),[]);title('Log Magnitude of Spectrum')

ifft_Bridge=abs(ifft2(fft_Bridge));   % inverse DFT and
                                        display
figure(2)
imshow(ifft_Bridge, [])
```

References

1 Deng, Y., Li, A., Chen, S., and Feng, D. (2018). Serviceability assessment for long-span suspension bridge based on deflection measurements. *Structural Control and Health Monitoring* 25: e2254.

2 Feng, D., Scarangello, T., Feng, M.Q., and Ye, Q. (2017). Cable tension force estimate using novel noncontact vision-based sensor. *Measurement* 99: 44–52.

3 Mao, J.-X., Wang, H., Feng, D.-M. et al. (2018). Investigation of dynamic properties of long-span cable-stayed bridges based on one-year monitoring data under normal operating condition. *Structural Control and Health Monitoring* 25: e2146.

4 Deng, Y., Li, A., and Feng, D. Fatigue performance investigation for hangers of suspension bridges based on site-specific vehicle loads. *Structural Health Monitoring* 18: 934–948.

5 Carden, E.P. and Fanning, P. (2004). Vibration based condition monitoring: a review. *Structural Health Monitoring* 3: 355–377.

6 Kim, J.T. and Stubbs, N. (2003). Crack detection in beam-type structures using frequency data. *Journal of Sound and Vibration* 259: 145–160.

7 Lee, J.J., Lee, J.W., Yi, J.H. et al. (2005). Neural networks-based damage detection for bridges considering errors in baseline finite element models. *Journal of Sound and Vibration* 280: 555–578.

8 Pandey, A.K., Biswas, M., and Samman, M.M. (1991). Damage detection from changes in curvature mode shapes. *Journal of Sound and Vibration* 145: 321–332.

9 Feng, M.Q., Kim, D.K., Yi, J.-H., and Chen, Y. (2004). Baseline models for bridge performance monitoring. *Journal of Engineering Mechanics* 130: 562–569.

10 Rahneshin, V. and Chierichetti, M. (2016). An integrated approach for non-periodic dynamic response prediction of complex structures: numerical and experimental analysis. *Journal of Sound and Vibration* 378: 38–55.

11 Xu, B., He, J., Rovekamp, R., and Dyke, S.J. (2012). Structural parameters and dynamic loading identification from incomplete measurements: approach and validation. *Mechanical Systems and Signal Processing* 28: 244–257.

12 Sun, H. and Betti, R. (2013). Simultaneous identification of structural parameters and dynamic input with incomplete output-only measurements. *Structural Control and Health Monitoring* 21: 868–889.

13 Feng, D., Sun, H., and Feng, M.Q. (2015). Simultaneous identification of bridge structural parameters and vehicle loads. *Computers & Structures* 157: 76–88.

14 Chen, Y. and Feng, M.Q. (2009). Structural health monitoring by recursive Bayesian filtering. *Journal of Engineering Mechanics* 135: 231–242.

15 Soyoz, S. and Feng, M.Q. (2008). Instantaneous damage detection of bridge structures and experimental verification. *Structural Control and Health Monitoring* 15: 958–973.

16 Feng, D. and Feng, M.Q. (2018). Computer vision for SHM of civil infrastructure: from dynamic response measurement to damage detection – a review. *Engineering Structures* 156: 105–117.

17 Poozesh, P., Baqersad, J., Niezrecki, C. et al. (2017). Large-area photogrammetry based testing of wind turbine blades. *Mechanical Systems and Signal Processing* 86 (Part B): 98–115.

18 Pratt, W.K. (2001). *Digital Image Processing: PIKS Inside*. Wiley.

19 Ye, X.W., Yi, T.-H., Dong, C.Z., and Liu, T. (2016). Vision-based structural displacement measurement: system performance evaluation and influence factor analysis. *Measurement* 88: 372–384.

20 Dworakowski, Z., Kohut, P., Gallina, A. et al. (2016). Vision-based algorithms for damage detection and localization in structural health monitoring. *Structural Control and Health Monitoring* 23: 35–50.

21 Pan, B., Tian, L., and Song, X. (2016). Real-time, non-contact and targetless measurement of vertical deflection of bridges using off-axis digital image correlation. *NDT&E International* 79: 73–80.

22 Awad, A.I. and Hassaballah, M. (2016). *Image Feature Detectors and Descriptors: Foundations and Applications*. Springer International Publishing.

23 Soh, Y., Qadir, M., Mehmood, A. et al. (2014). A feature area-based image registration. *International Journal of Computer Theory and Engineering* 6: 407–411.

24 Wang, Z., Kieu, H., Nguyen, H., and Le, M. (2015). Digital image correlation in experimental mechanics and image registration in computer vision: similarities, differences and complements. *Optics and Lasers in Engineering* 65: 18–27.

25 Park, S.W., Park, H.S., Kim, J.H., and Adeli, H. (2015). 3D displacement measurement model for health monitoring of structures using a motion capture system. *Measurement* 59: 352–362.

26 Wu, L.-J., Casciati, F., and Casciati, S. (2014). Dynamic testing of a laboratory model via vision-based sensing. *Engineering Structures* 60: 113–125.

27 Brownjohn, J., Hester, D., Xu, Y. et al. (2016). Viability of optical tracking systems for monitoring deformations of a long span bridge. Proceedings of EACS 2016 - 6th European Conference on Structural Control.

28 Feng, D., Feng, M.Q., Ozer, E., and Fukuda, Y. (2015). A vision-based sensor for noncontact structural displacement measurement. *Sensors* 15: 16557–16575.

29 Guizar-Sicairos, M., Thurman, S.T., and Fienup, J.R. (2008). Efficient subpixel image registration algorithms. *Optics Letters* 33: 156–158.

30 Ullah, F. and Kaneko, S. (2004). Using orientation codes for rotation-invariant template matching. *Pattern Recognition* 37: 201–209.

31 Feng, D. and Feng, M.Q. (2016). Vision-based multipoint displacement measurement for structural health monitoring. *Structural Control and Health Monitoring* 23: 876–890.

32 Fukuda, Y., Feng, M.Q., Narita, Y., Kaneko, S., and Tanaka, T. (2013). Vision-based displacement sensor for monitoring dynamic response using robust object search algorithm. *IEEE Sensors Journal* 13: 4725–4732.

33 Ullah, F., Kaneko, S., and Igarashi, S. (2001). Orientation code matching for robust object search (special issue on image recognition and understanding). *IEICE Transactions on Information and Systems* 84: 999–1006.

34 Zhang, Z. (2000). A flexible new technique for camera calibration. *IEEE Transactions on Pattern Analysis and Machine Intelligence* 22: 1330–1334.

35 Tian, Q. and Huhns, M.N. (1986). Algorithms for subpixel registration. *Computer Vision, Graphics, and Image Processing* 35: 220–233.

36 Bing, P., Hui-min, X., Bo-qin, X., and Fu-long, D. (2006). Performance of sub-pixel registration algorithms in digital image correlation. *Measurement Science and Technology* 17: 1615.

37 Moreu, F. and LaFave, J. (2012). Current research topics: Railroad bridges and structural engineering. NSEL Report Series: Report No. NSEL-032. University of Illinois at Urbana.

38 Shinozuka, M., Karmakar, D., Chaudhuri, S.R., and Lee, H. (2009). Verification of computer analysis models for suspension bridges. Technical report; Caltrans.

39 Mayer, L., Yanev, B., Olson, L.D., and Smyth, A. (2010). Monitoring of manhattan bridge for vertical and torsional performance with GPS and interferometric radar systems. Transportation Research Board 89th Annual Metting.

40 Chen, Y., Feng, M.Q., and Soyoz, S. (2008). Large-scale shake table test verification of bridge condition assessment methods. *Journal of Structural Engineering* 134: 1235–1245.

41 Frizzarin, M., Feng, M.Q., Franchetti, P. et al. (2010). Damage detection based on damping analysis of ambient vibration data. *Structural Control and Health Monitoring* 17: 368–385.

42 Gomez, H.C., Fanning, P.J., Feng, M.Q., and Lee, S. (2011). Testing and long-term monitoring of a curved concrete box girder bridge. *Engineering Structures* 33: 2861–2869.

43 Soyoz, S. and Feng, M.Q. (2009). Long-term monitoring and identification of bridge structural parameters. *Computer-Aided Civil and Infrastructure Engineering* 24: 82–92.

44 Chen, Y., Feng, M.Q., and Tan, C.-A. (2009). Bridge structural condition assessment based on vibration and traffic monitoring. *Journal of Engineering Mechanics* 135: 747–758.

45 Tan, C.A., Beyene Ashebo, D., Feng, M.Q., and Fukuda, Y. (2007). Integration of traffic information in the structural health monitoring of highway bridges. Proceedings of Sensors and Smart Structures Technologies for Civil, Mechanical, and Aerospace Systems.

46 Ribeiro, D., Calçada, R., Delgado, R. et al. (2012). Finite element model updating of a bowstring-arch railway bridge based on experimental modal parameters. *Engineering Structures* 40: 413–435.

47 Wiberg, J., Karoumi, R., and Pacoste, C. (2012). *Infrastructure Design, Signalling and Security in Railway*. InTech.

48 Feng, D. and Feng, M.Q. (2015). Model updating of railway bridge using in situ dynamic displacement measurement under trainloads. *Journal of Bridge Engineering* 20: 04015019.

49 Cheng, Y.S., Au, F.T.K., and Cheung, Y.K. (2001). Vibration of railway bridges under a moving train by using bridge-track-vehicle element. *Engineering Structures* 23: 1597–1606.

50 Liu, K., Reynders, E., De Roeck, G., and Lombaert, G. (2009). Experimental and numerical analysis of a composite bridge for high-speed trains. *Journal of Sound and Vibration* 320: 201–220.

51 Ju, S.-H., Lin, H.-T., and Huang, J.-Y. (2009). Dominant frequencies of train-induced vibrations. *Journal of Sound and Vibration* 319: 247–259.

52 Hollandsworth, P.E. and Busby, H.R. (1989). Impact force identification using the general inverse technique. *International Journal of Impact Engineering* 8: 315–322.

53 Inoue, H., Ikeda, N., Kishimoto, K. et al. (1995). Inverse analysis of the magnitude and direction of impact force. *JSME International Journal Series A: Mechanics and Material Engineering* 38: 84–91.

54 Wang, B.-T. and Chiu, C.-H. (2003). Determination of unknown impact force acting on a simply supported beam. *Mechanical Systems and Signal Processing* 17: 683–704.

55 Inoue, H., Harrigan, J.J., and Reid, S.R. (2001). Review of inverse analysis for indirect measurement of impact force. *Applied Mechanics Reviews* 54: 503–524.

56 Feng, D. and Feng, M.Q. (2017). Identification of structural stiffness and excitation forces in time domain using noncontact vision-based displacement measurement. *Journal of Sound and Vibration* 406: 15–28.

57 Mohan, A. and Poobal, S. (2018). Crack detection using image processing: a critical review and analysis. *Alexandria Engineering Journal* 57: 787–798.

58 Fang, Z. and Wang, J.Q. (2012). Practical formula for cable tension estimation by vibration method. *Journal of Bridge Engineering* 17: 161–164.

59 Feng, D., Mauch, C., Summerville, S., and Fernandez, O. (2018). Suspender replacement for a signature bridge. *Journal of Bridge Engineering* 23: 05018010.

60 D'Emilia, G., Razzè, L., and Zappa, E. (2013). Uncertainty analysis of high frequency image-based vibration measurements. *Measurement* 46: 2630–2637.

61 Stephen, G.A., Brownjohn, J.M.W., and Taylor, C.A. (1993). Measurements of static and dynamic displacement from visual monitoring of the Humber Bridge. *Engineering Structures* 15: 197–208.

62 Olaszek, P. (1999). Investigation of the dynamic characteristic of bridge structures using a computer vision method. *Measurement* 25: 227–236.

63 Wahbeh, A.M., John, P.C., and Sami, F.M. (2003). A vision-based approach for the direct measurement of displacements in vibrating systems. *Smart Materials and Structures* 12: 785.

64 Feng, M.Q., Fukuda, Y., Feng, D., and Mizuta, M. (2015). Nontarget vision sensor for remote measurement of bridge dynamic response. *Journal of Bridge Engineering* 20: 04015023.

65 Yoneyama, S. and Ueda, H. (2012). Bridge deflection measurement using digital image correlation with camera movement correction. *Materials Transactions* 53: 285–290.

66 Fukuda, Y., Feng, M.Q., and Shinozuka, M. (2010). Cost-effective vision-based system for monitoring dynamic response of civil engineering structures. *Structural Control and Health Monitoring* 17: 918–936.

67 Busca, G., Cigada, A., Mazzoleni, P., and Zappa, E. (2014). Vibration monitoring of multiple bridge points by means of a unique vision-based measuring system. *Experimental Mechanics* 54: 255–271.

68 Khuc, T. and Catbas, F.N. (2016). Completely contactless structural health monitoring of real-life structures using cameras and computer vision. *Structural Control and Health Monitoring* https://doi.org/10.1002/stc.852.

69 Yoon, H., Elanwar, H., Choi, H. et al. (2016). Target-free approach for vision-based structural system identification using consumer-grade cameras. *Structural Control and Health Monitoring* 23: 1405–1416. https://doi.org/10.1002/stc.850.

70 Cha, Y.-J., You, K., and Choi, W. (2016). Vision-based detection of loosened bolts using the Hough transform and support vector machines. *Automation in Construction* 71 (Part 2): 181–188.

71 Feng, D. and Feng, M.Q. (2017). Experimental validation of cost-effective vision-based structural health monitoring. *Mechanical Systems and Signal Processing* 88: 199–211.

72 Lee, J.-H., Ho, H.-N., Shinozuka, M., and Lee, J.-J. (2012). An advanced vision-based system for real-time displacement measurement of high-rise buildings. *Smart Materials and Structures* 21: 125019.

73 Ojio, T., Carey, C.H., OBrien, E.J. et al. (2016). Contactless bridge weigh-in-motion. *Journal of Bridge Engineering* 21: 04016032.

74 Hoskere, V., Park, J.-W., Yoon, H., and Spencer, B.F. (2019). Vision-based modal survey of civil infrastructure using unmanned aerial vehicles. *Journal of Structural Engineering* 145: 04019062.

75 Foroosh, H., Zerubia, J.B., and Berthod, M. (2002). Extension of phase correlation to subpixel registration. *IEEE Transactions on Image Processing* 11: 188–200.

76 Berenstein, C.A., Kanal, L.N., Lavine, D., and Olson, E.C. (1987). A geometric approach to subpixel registration accuracy. *Computer Vision, Graphics, and Image Processing* 40: 334–360.

77 Pan, B., Qian, K., Xie, H., and Asundi, A. (2009). Two-dimensional digital image correlation for in-plane displacement and strain measurement: a review. *Measurement Science and Technology* 20: 062001.

78 Debella-Gilo, M. and Kääb, A. (2011). Sub-pixel precision image matching for measuring surface displacements on mass movements using normalized cross-correlation. *Remote Sensing of Environment* 115: 130–142.

79 Ferrer, B., Mas, D., García-Santos, J.I., and Luzi, G. (2016). Parametric study of the errors obtained from the measurement of the oscillating movement of a bridge using image processing. *Journal of Nondestructive Evaluation* 35: 53.

80 Sutton, M., Yan, J.H., Tiwari, V., and Orteu, J.J. (2008). The effect of out-of-plane motion on 2D and 3D digital image correlation measurements. *Optics and Lasers in Engineering* 46: 746–757.

81 Baqersad, J., Poozesh, P., Niezrecki, C., and Avitabile, P. (2016). Photogrammetry and optical methods in structural dynamics – a review. *Mechanical Systems and Signal Processing* 86: 17–37.

82 Besnard, G., Hild, F., and Roux, S. (2006). "Finite-element" displacement fields analysis from digital images: application to Portevin–Le Châtelier bands. *Experimental Mechanics* 46: 789–803.

83 Roux, S. and Hild, F. (2006). Stress intensity factor measurements from digital image correlation: post-processing and integrated approaches. *International Journal of Fracture* 140: 141–157.

84 Haddadi, H. and Belhabib, S. (2008). Use of rigid-body motion for the investigation and estimation of the measurement errors related to digital image correlation technique. *Optics and Lasers in Engineering* 46: 185–196.

85 Kim, S.-W., Jeon, B.-G., Kim, N.-S., and Park, J.-C. (2013). Vision-based monitoring system for evaluating cable tensile forces on a cable-stayed bridge. *Structural Health Monitoring* 12: 440–456.

86 Luo, L., Feng, M.Q., Wu, Z.Y. (2018). Robust vision sensor for multi-point displacement monitoring of bridges in the field. *Engineering Structures*. 163: 255–66.

87 Lava, P., Van Paepegem, W., Coppieters, S. et al. (2013). Impact of lens distortions on strain measurements obtained with 2D digital image correlation. *Optics and Lasers in Engineering* 51: 576–584.

88 Pan, B., Yu, L., Wu, D., and Tang, L. (2013). Systematic errors in two-dimensional digital image correlation due to lens distortion. *Optics and Lasers in Engineering* 51: 140–147.

89 Yoneyama, S., Kikuta, H., Kitagawa, A., and Kitamura, K. (2006). Lens distortion correction for digital image correlation by measuring rigid body displacement. *Optical Engineering* 45: 023602–023609.

90 Ribeiro, D., Calçada, R., Ferreira, J., and Martins, T. (2014). Non-contact measurement of the dynamic displacement of railway bridges using an advanced video-based system. *Engineering Structures* 75: 164–180.

91 Lee, J.J. and Shinozuka, M. (2006). A vision-based system for remote sensing of bridge displacement. *NDT&E International* 39: 425–431.

92 Anantrasirichai, N., Achim, A., Kingsbury, N.G., and Bull, D.R. (2013). Atmospheric turbulence mitigation using complex wavelet-based fusion. *IEEE Transactions on Image Processing* 22: 2398–2408.

93 Feng, D. and Feng, M. (2016). Output-only damage detection using vehicle-induced displacement response and mode shape curvature index. *Structural Control and Health Monitoring* 23: 1088–1107.

94 Kohut, P., Holak, K., Uhl, T. et al. (2013). Monitoring of a civil structure's state based on noncontact measurements. *Structural Health Monitoring* 12: 411–429.

95 Chatzi, E.N. and Smyth, A.W. (2013). Particle filter scheme with mutation for the estimation of time-invariant parameters in structural health monitoring applications. *Structural Control and Health Monitoring* 20: 1081–1095.

96 Ho, H.-N., Lee, J.-H., Park, Y.-S., and Lee, J.-J. (2012). A synchronized multipoint vision-based system for displacement measurement of civil infrastructures. *The Scientific World Journal* 2012: 9.

97 Jeon, H., Bang, Y., and Myung, H. (2011). A paired visual servoing system for 6-DOF displacement measurement of structures. *Smart Materials and Structures* 20: 045019.

98 Hong, W., Zhang, J., Wu, G., and Wu, Z. (2015). Comprehensive comparison of macro-strain mode and displacement mode based on different sensing technologies. *Mechanical Systems and Signal Processing* 50–51: 563–579.

99 Nikfar, F. and Konstantinidis, D. (2016). Evaluation of vision-based measurements for shake-table testing of nonstructural components. *Journal of Computing in Civil Engineering* 31: 04016050.

100 Min, J.H., Gelo, N.J., and Jo, H. (2015). Non-contact and real-time dynamic displacement monitoring using smartphone technologies. *Journal of Life Cycle Reliability and Safety Engineering* 4: 40–51.

101 Ekin, O., Dongming, F., and Maria, Q.F. (2017). Hybrid motion sensing and experimental modal analysis using collocated smartphone camera and accelerometers. *Measurement Science and Technology* 28: 105903.

102 Warren, C., Niezrecki, C., Avitabile, P., and Pingle, P. (2011). Comparison of FRF measurements and mode shapes determined using optically image based, laser, and accelerometer measurements. *Mechanical Systems and Signal Processing* 25: 2191–2202.

103 Shariati, A. and Schumacher, T. (2016). Eulerian-based virtual visual sensors to measure dynamic displacements of structures. *Structural Control and Health Monitoring* 24: e1977.

104 Ye, X.W., Ni, Y.Q., Wai, T.T. et al. (2013). A vision-based system for dynamic displacement measurement of long-span bridges: algorithm and verification. *Smart Structures and Systems* 12: 363–379.

105 Wang, W., Mottershead, J.E., Siebert, T., and Pipino, A. (2012). Frequency response functions of shape features from full-field vibration measurements using digital image correlation. *Mechanical Systems and Signal Processing* 28: 333–347.

106 Wang, W., Mottershead, J.E., Ihle, A. et al. (2011). Finite element model updating from full-field vibration measurement using digital image correlation. *Journal of Sound and Vibration* 330: 1599–1620.

107 Cha, Y.J., Chen, J.G., and Büyüköztürk, O. (2017). Output-only computer vision based damage detection using phase-based optical flow and unscented Kalman filters. *Engineering Structures* 132: 300–313.

108 Oh, B.K., Hwang, J.W., Choi, S.W. et al. (2016). Dynamic displacements-based model updating with motion capture system. *Structural Control and Health Monitoring* 24: e1904.

109 Wang, W., Mottershead, J.E., and Mares, C. (2009). Vibration mode shape recognition using image processing. *Journal of Sound and Vibration* 326: 909–938.

110 Song, Y.-Z., Bowen, C.R., Kim, A.H. et al. (2014). Virtual visual sensors and their application in structural health monitoring. *Structural Health Monitoring* 13: 251–264.

111 Ji, Y. and Chang, C. (2008). Nontarget image-based technique for small cable vibration measurement. *Journal of Bridge Engineering* 13: 34–42.

112 Koch, C., Georgieva, K., Kasireddy, V. et al. (2015). A review on computer vision based defect detection and condition assessment of concrete and asphalt civil infrastructure. *Advanced Engineering Informatics* 29: 196–210.

113 Sony, S., Laventure, S., and Sadhu, A. A literature review of next-generation smart sensing technology in structural health monitoring. *Structural Control and Health Monitoring* 26: e2321.

114 Ham, Y., Han, K.K., Lin, J.J., and Golparvar-Fard, M. (2016). Visual monitoring of civil infrastructure systems via camera-equipped Unmanned Aerial Vehicles (UAVs): a review of related works. *Visualization in Engineering* 4: 1.

115 Kanellakis, C. and Nikolakopoulos, G. (2017). Survey on computer vision for UAVs: current developments and trends. *Journal of Intelligent and Robotic Systems* 87: 141–168.

116 Seo, J., Han, S., Lee, S., and Kim, H. (2015). Computer vision techniques for construction safety and health monitoring. *Advanced Engineering Informatics* 29: 239–251.

117 Ray, S.J. and Teizer, J. (2012). Real-time construction worker posture analysis for ergonomics training. *Advanced Engineering Informatics* 26: 439–455.

118 Han, S., Lee, S., and Peña-Mora, F. (2013). Vision-based detection of unsafe actions of a construction worker: case study of ladder climbing. *Journal of Computing in Civil Engineering* 27: 635–644.

119 Teizer, J. and Vela, P.A. (2009). Personnel tracking on construction sites using video cameras. *Advanced Engineering Informatics* 23: 452–462.

120 Azar, E.R. and McCabe, B. (2012). Automated visual recognition of dump trucks in construction videos. *Journal of Computing in Civil Engineering* 26: 769–781.

121 Chi, S. and Caldas, C.H. (2012). Image-based safety assessment: automated spatial safety risk identification of earthmoving and surface mining activities. *Journal of Construction Engineering and Management* 138: 341–351.

122 Brilakis, I., Park, M.-W., and Jog, G. (2011). Automated vision tracking of project related entities. *Advanced Engineering Informatics* 25: 713–724.

123 Chen, Y., Tan, C.-A., Feng, M.Q., and Fukuda, Y. (2006). *A Video Assisted Approach for Structural Health Monitoring of Highway Bridges Under Normal Traffic*. SPIE.

124 Gandhi, T., Chang, R., and Trivedi, M.M. (2007). Video and seismic sensor-based structural health monitoring: framework, algorithms, and implementation. *IEEE Transactions on Intelligent Transportation Systems* 8: 169–180.

125 Catbas, F.N., Zaurin, R., Gul, M., and Gokce, H.B. (2012). Sensor networks, computer imaging, and unit influence lines for structural health monitoring: case study for bridge load rating. *Journal of Bridge Engineering* 17: 662–670.

126 Feng, M.Q., Leung, R.Y., and Eckersley, C.M. (2020). Non-contact vehicle weigh-in-motion using computer vision. *Measurement* 153: 1–9.

127 Kim, J., Park, C.-S., and Min, K.-W. (2016). Fast vision-based wave height measurement for dynamic characterization of tuned liquid column dampers. *Measurement* 89: 189–196.

Index

a

acceleration   2, 3, 6, 8, 9, 64, 65, 92, 93, 100, 116, 120–130, 132–135, 140, 142, 187, 200, 203
accelerometer   3, 6, 9, 47, 59–61, 63, 64, 81, 82, 90–94, 98–100, 103, 104, 111, 112, 139, 171, 175, 178, 199–202, 204
ambient vibration test   101
analytical model   7, 8, 89, 92, 93, 98, 100–102, 104, 115, 116, 123, 130, 137, 141, 157, 159, 202, 207
anti-aliasing filter   198
artificial intelligence   204, 209
artificial neural network   3, 193
artificial target   8, 24, 27, 43–51, 53, 54, 63, 64, 69, 71, 72, 77, 79–82, 92, 93, 103–105, 192, 199–201

b

background disturbance   51, 88, 198
background image disturbance   8, 43
background occlusion   37
baseline model   101
basic principle   1
Bayesian filtering   2, 207
beam structure   8, 59, 97, 100, 108, 110–113, 139, 140, 154, 159, 203

b

bilinear interpolation   37
binary image   211
blind source separation   2
boundary conditions   101, 172, 174, 176
bridge maintenance   69
Bronx-Whitestone Bridge   9, 172, 184, 187, 204

c

cable-stayed bridge   171
cable-supported bridge   9, 187, 203
cable-supported canopy   175
cable-supported structure   171, 174, 189
cable tension force   171–174, 182, 187
cable vibration   9, 171, 174, 176, 178, 182, 196, 204
camera calibration   23–25, 40, 197
camera vibration   79, 80, 86, 174, 196, 201
central difference method   102, 108
charge-coupled device (CCD)   11, 196, 198, 214
checkerboard   24
civil engineering   7, 56, 191, 194, 198, 200, 204
civil infrastructure   1, 176, 202, 204, 208, 209

Computer Vision for Structural Dynamics and Health Monitoring, First Edition.
Dongming Feng and Maria Q. Feng.
© 2021 John Wiley & Sons Ltd.
This Work is a co-publication between John Wiley & Sons Ltd and ASME Press.
Companion website: www.wiley.com/go/feng/structuralhealthmonitoring

complementary metal-oxide semiconductor (CMOS)   11, 12, 44, 59, 69, 92, 176, 185, 198

computer vision   3–9, 11, 12, 15, 18, 21, 24, 28, 40, 43, 44, 46, 56, 62, 64, 66, 67, 69, 80, 81, 84, 87, 88, 91, 92, 94, 98, 103, 104, 108, 109, 111,–113, 115, 116, 121, 136, 137, 140, 147, 149, 154, 155, 157, 159, 171–174, 176, 178, 180, 185, 187, 189, 191–193, 195, 196, 198–209

computer vision-based sensor   3, 4, 7

condition assessment   1, 7, 171, 191, 208, 209

construction control   171, 172, 176

contact-type sensor   3–5, 67, 69, 80, 81, 91, 115, 171, 178, 187

conventional   1, 3–5, 7–9, 12, 16, 37, 51, 56, 64, 67, 69, 80, 81, 87, 88, 91, 94, 100, 111, 113, 121, 149, 171, 174, 178, 187, 198, 199, 201, 202, 205

convergence   3, 8, 103, 144–147, 158

coordinate conversion   7, 11, 22, 24, 41, 197

crack detection   205

cross-correlation   8, 11, 16, 18, 28, 29, 46, 193, 195, 204

d

damage detection   2, 7, 89, 91, 108, 110–113, 202–204, 206, 208, 209

damage indicator   2

damping ratio   142, 145–147, 150, 151, 154, 156, 208

data acquisition   1, 3, 77, 98, 171

data transmission   3

daytime   8, 71, 72, 77

decision-making   1, 209

defect detection   7

degradation   1, 184, 208

degrees of freedom (DOF)   4, 89, 102, 118–120

descriptors   20, 21, 203

detection of damage   1

deteriorate   1, 67, 77, 79

digital camera   4, 11, 198, 205

digital image processing   9, 211

digital video camera   11

digital video image   4, 109

dim light   8, 43, 51, 54, 55, 80

discrepancy   2, 93, 101, 104, 122, 139

discrete Fourier transform (DFT)   9, 28, 217

displacement measurement   4, 7, 11, 15, 16, 24, 29, 33, 40, 80, 84, 85, 87, 91, 93, 103, 111, 113, 116, 125, 137, 140, 142, 146, 149, 155, 157, 158, 173, 191–197, 199–201, 208

displacement response   3, 8, 84, 115, 116, 124–127, 135–137, 140, 145, 147, 149–151, 156, 158, 191, 200, 203

dissimilarity   34

dominant frequencies   116, 131

dynamic characteristic   89, 130

dynamic effects   116, 135

dynamic response   1, 8, 63, 66, 89, 118–120, 122–124, 135, 136, 208

e

edge detection   9, 216

effectiveness   2, 3, 82, 146, 196, 207

efficiency   18, 115, 171, 194, 195

Eigensystem Realization Algorithm (ERA)   92

eighting factors   102

El Centro Earthquake ground motion   43

energy dissipation   142

engineering structures   3, 7, 191, 198, 199, 204

environmental conditions   8, 43, 51, 67, 88, 198

Euclidean distance   20

excessive deflections   67

excitation   1, 2, 8, 44, 46, 51, 53, 58, 63, 87, 90–92, 94, 104, 115, 116, 130, 132, 134–136, 139–142, 144–147, 157–159, 176, 202, 203, 207

experimental modal analysis   60, 89,
    91–94, 100, 101, 103, 104, 108, 110, 112,
    113, 115, 136, 201, 202
experimental setup   67
extended Kalman filter   2
extent of damage   3
extrinsic parameter   23, 24, 26

f

feature-based template matching
    20, 28
feature detection   20, 216
field of view (FOV)   85, 86, 88, 147, 157,
    192, 193, 196, 199, 200
field tests   26, 27, 43, 63, 69–71, 79, 80,
    117, 191, 196, 200
finite element model   8, 89, 91, 101, 116,
    118, 120, 121, 127, 137, 139, 144, 202, 203
fixed connection   176
flexibility   1, 5, 12, 39, 80, 194
flexural rigidity   146, 172
focal length   12, 24, 25, 27, 40, 59, 174,
    176, 186, 197
force identification   139, 142
Fourier transform   9, 28, 29, 174, 217–219
frame rate   11, 12, 39, 81, 84, 174, 195, 198
frame structure   8, 46, 47, 49, 51, 56,
    91–94, 102–104, 112, 113, 202, 203
free vibration   59, 62, 100, 136, 178
frequency-domain   1, 101, 104, 113,
    116, 200
frequency-domain decomposition   1–2
frequency response function   1, 202

g

Gauss-Newton optimization
    algorithm   102, 147
geometrical distortion   40
global shutter   176, 198
GPS   4, 5, 84, 201
gradient   20, 28, 33, 41, 55, 56, 88, 127,
    193, 196, 199
gray image   33

grayscale image   211–213
ground motion   43, 57, 58, 80, 84, 140,
    159, 196

h

hammer impact test   97, 98, 109, 110, 154
Hankel matrix   143, 145
heat haze   79, 80, 198, 199, 208
high-contrast artificial targets   8, 43, 48,
    92, 192
high-resolution   47, 48, 113, 199, 201
highway bridge   2, 8, 43, 66, 90, 101, 118,
    200, 206, 207

i

identify   2, 3, 8, 16, 97, 116, 126, 127, 129,
    136, 137, 139–141, 155, 157, 174, 201,
    203, 204, 206
ill conditions   37
illumination   16, 20, 33, 37, 51, 75, 82, 88,
    195, 198–201, 209
image distortion   80, 196, 199, 208
image intensity   28, 41, 55, 88, 216
image plane   24–27, 61
image quality   51, 198
incomplete   2, 101
initial guess   3, 145
input forces   2, 139, 140, 144, 146, 149, 159
input-output   2, 122, 142
in situ vibration measurement   101, 116
intelligent infrastructure
    maintenance   209
interferometric radar system   4, 84
intrinsic parameter   23, 24

j

Jacobian matrix   102, 145

l

labor-intensive   1, 3, 204
large-scale structure   2, 3, 86, 205, 208
laser displacement sensors (LDSs)   47, 92,
    94, 154, 155

laser vibrometer   4–6, 67, 200, 201
least-squares estimation   2
lens distortion   40, 197
light condition   80, 82
linear variable differential transformer
   (LVDT)   4, 6, 44, 46, 57, 58, 63, 67, 69,
   72, 75, 77, 80, 115, 121, 200, 201
logarithmic decrement method   150,
   151, 156
long-term material deterioration   101
low-contrast natural targets   8, 48, 92, 201

m
machine learning   7
Manhattan Bridge   43, 82, 84, 86, 201
measurement accuracy   4, 5, 43, 47, 51,
   63, 77, 79, 80, 88, 102, 155, 191, 193,
   196–198, 201, 208
measurement distance   4, 6, 72, 79,
   80, 199
measurement resolution   49, 86, 147, 174,
   192, 193, 195, 200
metric   16, 125
misalignment   26
modal   1, 2, 7, 8, 59, 60, 89–113, 115, 116,
   121, 122, 136, 199, 201–204, 207
modal analysis   1, 7, 8, 60, 89, 91–94, 97,
   98, 100–104, 108, 110, 112, 113, 115, 116,
   136, 199, 201, 202, 204
modal flexibility   1
modal properties   1, 89, 91, 92, 94, 97, 103,
   116, 122, 202
modal strain energy   1
model updating   8, 89–91, 101, 104, 108,
   112, 115–117, 121, 122, 127, 136, 137,
   203, 204
mode shape   1, 2, 7, 89, 91–94, 97, 98,
   100–103, 108–110, 112, 122, 193, 201–204
mode shape curvature   2, 91, 108, 110,
   203, 204
monitoring   1, 3, 4, 6, 7, 46, 80, 88, 90, 137,
   171, 172, 175, 189, 191, 195, 199, 201,
   202, 204–209

Monte Carlo analysis   147
moving load model   118
moving mass model   118
moving spring-damper system model   118
moving trainloads   69, 136
multipoint measurement   12, 43, 84, 86,
   113, 192, 201, 202

n
natural frequency   1, 2, 7, 64, 89, 91–93,
   97, 98, 101–103, 112, 116, 122, 132, 136,
   142, 156, 171, 172, 174, 180, 187,
   198, 202
natural target   8, 43, 44, 46, 48, 49, 51, 53,
   69, 75, 79, 92, 93, 103, 112, 192, 200, 201
noise removal   9, 214–216
noncontact   4–6, 9, 67, 91, 94, 100, 113,
   115, 157, 171, 175, 182, 187, 204
noncontact type   4
nondestructive testing   1
normalized root mean squared error
   (NRMSE)   45, 98, 155
Nyquist theorem   198

o
objective function   2, 101, 103, 122–130,
   139, 141, 144, 145, 147
online   2, 3, 39
on-structure sensor   1, 3, 51, 87, 88, 91,
   100, 113
optical axis   25–27, 61–63, 88, 98, 197
optical flow   202, 204
optimization   2, 3, 8, 24, 101–104, 113,
   116, 122, 123, 127, 136, 137, 139–141,
   144, 147, 156, 158, 193, 203
ordinary least squares solution   143
orientation angle   33, 37
orientation code matching (OCM)   8, 11,
   28, 33–35, 37–41, 44, 46, 48–56, 69, 80,
   82, 87, 88, 92, 93, 103, 108, 109, 171, 173,
   174, 192, 195, 199, 201
outdoor   8, 26, 43, 55, 88, 195,
   198–200, 208

output-only   2, 8, 9, 140, 141, 144, 149,
    158, 159, 202, 206, 207

p

parameter identification   7, 100, 113, 141
partial target occlusion   51, 198
partial template occlusion   8, 43, 51, 88
particle filter   2
pedestrian bridge   8, 43, 63, 200
perpendicularity   25–27
pixel coordinate   20, 25, 38, 73, 97,
    193, 227
pixel-level   15, 16, 26, 60, 82, 174, 202
point-wise   3, 5
post-processing   12, 39, 51, 80, 194
practical calibration method   23, 25, 40,
    59–61, 197
prerequisite   2, 25
probabilistic   2
projective distortion   62, 197
projective transformation   24

q

quantitative assessment   1

r

railway bridge   8, 43, 67, 75, 79, 115, 116,
    118–120, 122, 127, 130, 135, 136, 139,
    200, 203
random decrement   2
random noise   94, 147
Rayleigh damping   120, 123, 142, 151, 156
region of interest (ROI)   15, 174, 195
remaining life   101
residual function   102, 145
resolution   3, 4, 11, 12, 27–29, 33–39,
    47–49, 81, 86, 88, 100, 108–113, 147, 157,
    174, 176, 185, 192–195, 200–203, 205, 206
risk assessment   3
robustness   8, 37, 43, 51, 80–82, 87, 88, 91,
    146, 147, 149, 150, 158, 191, 192, 200, 201
rolling shutter   176, 198

s

safety   1, 4, 67, 174, 175, 191, 201, 206, 209
sampling rate   44, 47, 57, 59, 67, 92, 109,
    176, 186, 198
scaling factor   15, 22, 23, 25–27, 40, 44,
    47–49, 60–63, 71, 81, 82, 84, 86, 88, 98,
    174, 187, 193, 197, 202
scaling factor determination   40
seismic shaking table   43, 56, 208
sensitivity   8, 102, 109, 116, 121, 123–127,
    136, 137, 192, 195, 203
shading   33, 198
shaking table   3, 43, 44, 46, 51, 56, 58, 91,
    92, 193, 199, 200, 208
simultaneous identification   2, 8, 140, 144,
    149, 150, 154, 159
stability   3, 67, 115, 199
state-of-the-art   9, 69, 191, 205
state-space representation   142, 145, 158
stationary reference point   4, 5, 44, 67, 69,
    80, 81, 86, 92, 115, 121
stereovision   194, 202
stiffness   3, 8, 67, 89, 93, 101–104, 108,
    109, 113, 116, 120–129, 134, 136, 137,
    139, 142, 145–159, 172, 202–204
stochastic subspace identification   2
structural damage   1, 89, 101, 108, 109,
    113, 202–204
structural dynamics   1, 3, 7, 89, 200,
    208, 209
structural health monitoring (SHM)
    1–3, 5, 7–9, 90, 91, 191, 192, 199, 201,
    204, 209
structural integrity   1, 7
structural parameters   2, 8, 101, 104, 108,
    120, 139–141, 143–146, 157, 158, 206
structural response   2, 58, 90, 91, 104, 115,
    122, 139, 144, 157, 159, 199, 203
structural system   1, 122, 141, 201
subpixel   28, 29, 41, 47–49, 88, 108, 109,
    171, 174, 193, 194, 203, 207
sum of squared differences (SSD)   16

surface reflection 37
suspension bridge 8, 43, 81, 82, 86, 171, 184
system identification 1, 2, 141, 202

t

target panel 27, 40, 44, 46, 57, 58, 63, 67, 69, 71, 72, 79–82, 201
taut string theory 172
telescopic lens 58, 67
template matching 8, 11, 15, 16, 18, 20, 28, 29, 33, 37, 39–41, 46–49, 51, 53–56, 80, 87, 191–193, 195, 198, 199
temporal aliasing 198
Tikhonov regularization 143
time-consuming 1, 3, 4, 40, 171, 187, 204
time-domain 1–3, 8, 116, 122, 127, 136, 139, 141, 158, 159, 203
time-invariant 144
time synchronization 3, 195
time-varying system 116
train-track-bridge interaction 116, 118, 119, 135
triangular geometry 26
true color image 211

u

uncertainties 2, 27, 63, 101, 121, 123, 145, 151, 174, 176, 182, 197
unscented Kalman filter 2, 202
upsampled cross-correlation (UCC) 8, 11, 28, 29, 39–41, 46, 48–57, 87, 88, 195
upsampling 29, 48, 193

v

vehicle loads 2, 66, 200, 206
vertical displacement 67, 69, 81, 84–86, 98, 117, 125, 196, 201
vibration method 9, 171, 172, 175, 182, 187
video images 4, 5, 7, 11, 12, 16, 22, 28, 39, 40, 44, 47, 48, 59, 60, 92, 109, 137, 194, 195, 207
Vincent Thomas Bridge 43, 81, 201
vision-based displacement sensor 5, 15, 75, 191, 195, 199
vision-based sensing 7, 208
vision sensor 3–9, 11, 12, 15, 16, 29, 38–41, 43–58, 62–70, 72, 75, 77, 79–82, 84, 86–88, 91–94, 98–100, 103, 108, 109, 111–113, 115–118, 121, 136, 137, 140, 147, 149, 154, 155, 157, 159, 171–176, 178, 180, 182, 185, 187, 189, 191–204, 207, 208
vision sensor system 7, 8, 11, 12, 15, 40, 41, 43, 46, 47, 64, 67, 69, 81, 84, 87, 98, 100, 117, 171, 174, 176, 185, 194, 195, 200, 201
visual inspection 1, 191, 204–206, 209

w

wired sensors 3
wireless sensing 3

z

zoom lens 5, 11, 24, 27, 79, 80, 154, 173, 196, 197